Lecture Notes
in Business Information Processing   **219**

Series Editors

Wil van der Aalst
  Eindhoven Technical University, Eindhoven, The Netherlands
John Mylopoulos
  University of Trento, Povo, Italy
Michael Rosemann
  Queensland University of Technology, Brisbane, QLD, Australia
Michael J. Shaw
  University of Illinois, Urbana-Champaign, IL, USA
Clemens Szyperski
  Microsoft Research, Redmond, WA, USA

More information about this series at http://www.springer.com/series/7911

Joonsoo Bae · Suriadi Suriadi
Lijie Wen (Eds.)

Asia Pacific Business Process Management

Third Asia Pacific Conference, AP-BPM 2015
Busan, South Korea, June 24–26, 2015
Proceedings

 Springer

Editors
Joonsoo Bae
Chonbuk National University
Jeonju
Korea, Republic of (South Korea)

Lijie Wen
Tsinghua University
Beijing
China

Suriadi Suriadi
Massey University
Albany
New Zealand

ISSN 1865-1348   ISSN 1865-1356 (electronic)
Lecture Notes in Business Information Processing
ISBN 978-3-319-19508-7   ISBN 978-3-319-19509-4 (eBook)
DOI 10.1007/978-3-319-19509-4

Library of Congress Control Number: 2015939678

Springer Cham Heidelberg New York Dordrecht London

Printed on acid-free paper

Springer International Publishing AG Switzerland is part of Springer Science+Business Media
(www.springer.com)

Preface

This volume collects the proceedings of the Asia-Pacific Conference on Business Process Management 2015 (AP-BPM 2015) held in Busan, Korea, during June 24–26, 2015. AP-BPM 2015 was the third edition of the conference for researchers and practitioners in the field of business process management (BPM) in the Asia-Pacific region. It aims to provide a high-quality forum for researchers and practitioners to exchange research findings and ideas on BPM technologies and practices that are highly relevant to the Asia-Pacific region. At the same time, the conference committee welcomes submissions on relevant topics from all over the world. A key objective of the conference is to build a bridge between actual industrial requirements and leading-edge research outcomes in the Asia-Pacific region. The theme of AP-BPM 2015 was "Business Processes Meet Big Data."

As the third edition in this conference series, AP-BPM 2015 attracted quite a few submissions: 37 (qualified) submissions. These submissions reported on up-to-date research findings of scholars from ten countries (Australia, China, Denmark, Indonesia, Italy, Korea, Malaysia, New Zealand, Taiwan, Turkey, and USA). After each submission was reviewed by at least three Program Committee members, 12 full papers and two short papers were accepted for publication in this volume of conference proceedings (i.e., 32.4 % acceptance rate for full papers and 5.4 % for short papers). These 14 papers cover various topics that can be categorized under five main research focuses in BPM, including "Algorithms for Process Analysis" (two papers), "Process Mining" (three papers), "Resources Allocation Strategies" (two papers), "Advancement in Workow Technologies" (four papers), and "Emerging Topics in BPM" (three papers).

We would like to thank the Program Committee members for their thorough reviews and discussions of the submitted papers. We express our gratitude to other conference committees as well, especially to the general chair, Hyerim Bae, and the Steering Committee for their valuable guidance, to the organizing chair, Minseok Song, and other staff at Ulsan National Institute of Science and Technology (UNIST), Korea, for their attentive preparations for this conference, to the publicity chairs, Raffaele Conforti, Zaiwen Feng, and Bernardo Nugroho Yahya, for their efforts in publishing conference updates and promoting the conference in the region, to the Conference Advisory Committee for their kind advice, to the industrial session chairs, Jae-Yoon Jung and Sang-Hoon Lee for their great efforts in preparing the industrial session, and to the Process Mining Competition chair, KwangSup Shin, for organizing the competition event. Last but not least, we are thankful to the authors of the submissions, the presenters, and all the other conference participants, because the conference could not be held without their contributions and interest.

April 2015

Joonsoo Bae
Suriadi Suriadi
Lijie Wen

Organization

APBPM 2015 was organized by the Ulsan National Institute of Science and Technology, Korea.

General Chair

Hyerim Bae   Pusan National University, Korea

Steering Committee

Hyerim Bae   Pusan National University, Korea
Arthur ter Hofstede   Queensland University of Technology, Australia
Jianmin Wang   Tsinghua University, China

Program Co-chairs

Joonsoo Bae   Chonbuk National University, Korea
Suriadi Suriadi   Massey University, New Zealand
Lijie Wen   Tsinghua University, China

Organizing Committee Chair

Minseok Song   Ulsan National Institute of Science and Technology, Korea

Publicity Chairs

Raffaele Conforti   Queensland University of Technology, Australia
Zaiwen Feng   Wuhan University, China
Bernardo Nugroho Yahya   Hankuk University of Foreign Studies, Korea

Conference Advisory Committee

Chia-Yu Hsu   Yuan Ze University, Taiwan
Iwan Vanany   Sepuluh Nopember Institute of Technology, Indonesia

Industrial Session Chairs

Jae-Yoon Jung   Kyung Hee University, Korea
Sang-Hoon Lee   Oracle, Korea

Process Mining Competition Chair

KwangSup Shin   Incheon University, Korea

Program Committee

Saiful Akbar	Bandung Institute of Technology, Indonesia
Majed Al-Mashar	King Saud University, Saudi Arabia
Joonsoo Bae	Chonbuk National University, Korea
Hyerim Bae	Pusan National University, Korea
Jian Cao	Shanghai Jiao Tong University, China
Namwook Cho	Seoul National University of Science and Technology, Korea
Lizhen Cui	Shandong University, China
Zaiwen Feng	Wuhan University, China
Jidong Ge	Nanjing University, China
Arthur ter Hofstede	Queensland University of Technology, Australia
Jae-Yoon Jung	Kyung Hee University, Korea
Dongsoo Kim	Soongsil University, Korea
Kwanghoon Kim	Kyonggi University, Korea
Minsoo Kim	Pukyong National University, Korea
Raymond Lau	City University of Hong Kong, SAR, China
Ying Li	Zhejiang University, China
Xiao Liu	East China Normal University, China
Chun Ouyang	Queensland University of Technology, Australia
Helen (Hye-Young) Paik	University of New South Wales, Australia
Artem Polyvyanyy	Queensland University of Technology, Australia
Punnamee Sachakamol	Kasetsart University, Thailand
Shazia Sadiq	University of Queensland, Australia
Lawrence Si	University of Macau, SAR, China
Minseok Song	Ulsan National Institute of Science and Technology, Korea
Jianwen Su	University of California at Santa Barbara, USA
Suriadi Suriadi	Massey University, New Zealand
Jianmin Wang	Tsinghua University, China
Ingo Weber	NICTA, Australia
Lijie Wen	Tsinghua University, China
Raymond Wong	University of New South Wales, Australia
Jei-Zheng Wu	Soochow University, Taiwan
Moe Thandar Wynn	Queensland University of Technology, Australia
Bernardo Nugroho Yahya	Hankuk University of Foreign Studies, Korea
Zhiqiang Yan	Capital University of Economics and Business, China
Jianwei Yin	Zhejiang University, China
Sira Yongchareon	Unitec Institute of Technology, New Zealand
Yang Yu	Sun Yat-Sen University, China

Qingtian Zeng   Shandong University of Science and Technology,
                             China
Liang Zhang   Fudan University, China
Yang Zhang   Beijing University of Posts and Telecommunications,
                             China

Contents

Process Mining

Emerging Topics in BPM

Advancement in Workflow Technologies

Realisation of Cost-Informed Process Support Within the YAWL Workflow Environment

M. Adams[1]([✉]), M.T. Wynn[1], C. Ouyang[1], and A.H.M. ter Hofstede[1,2]

[1] Queensland University of Technology, Brisbane, Australia
{mj.adams,m.wynn,c.ouyang,a.terhofstede}@qut.edu.au
[2] Eindhoven University of Technology, Eindhoven, The Netherlands

Abstract. Organisations are always focussed on ensuring that their business operations are performed in the most cost-effective manner, and that processes are responsive to ever-changing cost pressures. In many organisations, however, strategic cost-based decisions at the managerial level are not directly or quickly translatable to process-level operational support. A primary reason for this disconnect is the limited system-based support for cost-informed decisions at the process-operational level *in real time*. In this paper, we describe the different ways in which a workflow management system can support process-related decisions, guided by cost-informed considerations at the operational level, during execution. As a result, cost information is elevated from its non-functional attribute role to a first-class, fully functional process perspective. The paper defines success criteria that a WfMS should meet to provide such support, and discusses a reference implementation within the YAWL workflow environment that demonstrates how the various types of cost-informed decision rules are supported, using an illustrative example.

Keywords: Cost-informed process enactment · Workflow management · Yet Another Workflow Language (YAWL)

1 Introduction

Organisations are constantly seeking cost efficiencies in their day-to-day operations, and are eager to have cost-based considerations applied to their business processes in a timely manner. In most organisations, however, there is a temporal disconnect between the making of strategic cost-based decisions and having them applied as business rules in automated business processes. Our observation is that most Workflow Management Systems (WfMSs) offer no support for cost considerations beyond the use of generic non-functional attributes or basic *a posteriori* reporting. Detailed *real-time* cost information is typically unavailable during process execution and, as a result, system-based decisions cannot be made within the context of a process instance while it is active, nor can it be used for live monitoring or other operational decision support.

Cost is often described as a non-functional requirement (NFR) for a software system, in the same class as requirements such as maintainability, usability,

© Springer International Publishing Switzerland 2015
J. Bae et al. (Eds.): AP-BPM 2015, LNBIP 219, pp. 3–18, 2015.
DOI: 10.1007/978-3-319-19509-4_1

reliability, traceability, quality or safety [1]. However, care must be taken to distinguish between the implementation cost requirements of a host process-aware system and the impact of a cost-informed perspective on each of the processes it subsequently executes and manages. A WfMS is a particular kind of software system that executes process models that in turn are very closely linked with cost implications at all scopes of a business process: individual activities, resources, process instances and execution sets. Also, the dynamic and sometimes ad-hoc nature of strategic cost-related decisions makes their timely application of utmost importance at both the design time and runtime phases of a process. There are potentially significant cost savings that an organisation can achieve through WfMS support for system-based, real-time, cost-informed decisions.

A short research paper presenting a conceptual framework that defines how WfMSs can provide sophisticated support for strategic cost-informed operational decisions was published in [2]. This paper extends the earlier work by providing a detailed discussion of the realisation of the cost-informed operational support within the well-known open-source WfMS system environment YAWL [3], illustrated with a familiar example. The remainder of the paper is organised as follows. Section 2 provides a review of the conceptual framework. Section 3 defines the success criteria that need to be satisfied before a WfMS can be considered cost-informed. Section 4 discusses the implementation of a *Cost Service* within YAWL. Section 5 describes an example that illustrates the implementation. Section 6 presents related work and Sect. 7 concludes the paper.

2 A Framework for Cost-Informed Decisions

In order for WfMSs to support cost-informed decision making during process execution, a number of key requirements must be fulfilled. Firstly, a WfMS must be 'cost-aware', i.e. there must be support for the ability to link cost information (e.g. cost rates, cost functions) to different elements in a process (e.g. activities, case attributes and durations) and to resources (e.g. names, roles or experience levels). Secondly, a cost-informed WfMS should provide the ability to specify cost-informed control flow definitions and resource allocation rules *at design time* and to provide support for system-based decisions and system-supported user decisions *at runtime*. Thirdly, a cost-informed WfMS should provide the ability to compute processing costs of a case *at runtime* and to provide a mechanism for a detailed cost analysis *at post-execution time*.

Figure 1 depicts a conceptual framework which describes (1) *data input*, i.e. the information requirements to enact actions that can be undertaken by or with a WfMS to support cost-informed decision making, (2) the *actions* that can be taken at the levels of process, activity, and resource (work distribution), and (3) the *cost-informed support* that is delivered, either through decisions by the WfMS itself or people using its support.

To support cost-informed actions with a WfMS, we require an *executable process model* and a *cost model* that describes all cost rates/data associated with activities within the process. Cost data could be as simple or as complex as an

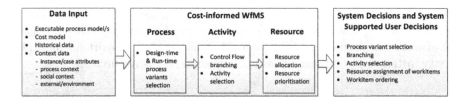

Data Input	Cost-informed WfMS			System Decisions and System Supported User Decisions

• Executable process model/s
• Cost model
• Historical data
• Context data
  - instance/case attributes
  - process context
  - social context
  - external/environment
	Process
	
	• Design-time & Run-time process variants selection
		Activity
		
		• Control Flow branching
		• Activity selection
			Resource
			
			• Resource allocation
			• Resource prioritisation
				• Process variant selection
				• Branching
				• Activity selection
				• Resource assignment of workitems
				• Workitem ordering

Fig. 1. A framework supporting cost-informed process execution with a WfMS.

organisation requires. For instance, it could be a variable cost that describes the hourly rate of a resource, but it could also be a function to compute the cost of an activity based on a combination of the resource assigned at runtime, the type of case being executed and the duration of the executed activity. Cost information, together with *historical data* as stored in a so-called *process log* regarding past executions, can be used to determine the cost of process executions, as illustrated in our earlier work [4]. Since a business process is always executed in a particular context, we also adopt the four levels of *context data* described in [5]: case attributes, process context, social context, and the environment.

All *cost-informed actions* supported by a WfMS are informed by data input and governed by strategic cost optimisation goals. These actions can be classified into three levels: process, activity, and resource.

Process. The *process* level is concerned with making *cost-informed process selection* decisions based on cost information of processes or process variants at design time or at runtime. This may involve the selection among different processes or selection among different process variants (which are created from individual processes during the execution phase). It should be possible to assign a (whole) process or process variant to a certain resource team for execution (i.e. outsourcing) based on the cost profile and a certain set of cost-based selection rules. For instance, a WfMS may make an automated selection of a process variant based on predefined cost-based selection rules. Alternatively, the WfMS may provide an administrator with cost profiles of different process variants and s/he can make an informed selection.

Activity. For cases that have been started under the control of a WMFS, it is necessary to decide at certain points about the *activity* (or activities) to be executed next. In its coordination capability, a WfMS may decide on which activity instances are enabled in a specific case, based on the branching conditions specified in the underlying process. As such, the WfMS can make a cost-informed decision based on a pre-defined business rule to enable/start an activity. In a similar manner, a WfMS could also skip, suspend or cancel activity instances, among other actions, based on that cost information. Or, an administrator can do so based on the cost information presented to them.

Resource. For activity instances that need to be carried out by a user, both "push" and "pull" patterns of activity-resource assignment [6] should be supported. That is, a WfMS would need to support cost-informed resource allocation rules for selecting a resource to offer, allocate, and/or start an activity instance and to support a user to make such cost-informed activity selection decisions. Finally, a WfMS should provide support for ordering and prioritisation of work based on cost information. That means that support for multiple activity instances to be ranked based on expected cost is required.

3   Implementation Success Criteria

For a WfMS to be capable of providing full cost-informed support across all three levels (process, activity and resource), it is essential that the following key success criteria be satisfied.

1. **Associating Processes with Cost Data and Cost-based Rules.** Processes, activities and resources must have their relevant cost rates specified, stored and accessible during execution. Example values would include salary and incidental costs for human resources, the costs of required materials and consumables, and fixed costs associated with activity enactments, rentals, depreciation, etc. Definitions of such costs would normally be acquired at process design time and either incorporated directly into the process specification or linked to it via an external file or database.
2. **Dynamic Cost Calculations at Runtime.** The dynamic runtime calculation of the cost of each process instance and its components is required so that real-time system- and human-based decisions can be applied. Such calculations would fall into the following categories:
   - *time-based*, e.g. salary costs for the period a human resource spends on performing an activity, or timed charges for interaction with an external service, or the cost of insurance for the duration of an activity;
   - *usage-based*, e.g. forklift hire, or the use of an MRI machine, or payment of a set fee for an expert witness;
   - *measurement-based*, e.g. per tonne costs of a raw material, or per millilitre of a pharmaceutical, or per kilowatt-hour of an energy supply;
   - *invocation-based*, e.g. costs involved in retooling an assembly line for a product run;
   - *a fixed cost*, e.g. an overhead cost of commencing an activity, or a building approval application fee;
   - *a combination of the above*, e.g. a truck rental may involve an initial hire cost plus a fee per kilometre plus a fuel cost, or a usage-based fee for a machine hire may also involve a time-based insurance fee.
3. **Cost Data Logging and Analysis.** All calculated costs for each process instance and component activities must be incorporated into the process event logs so that they can be reviewed by management during and post execution, and used to perform extrapolated calculations over archived data in real time as required for the next criterion.
4. **Support for Cost-informed Decisions.** A cost-informed WfMS must support the ability to use the calculated costs of the current process instance and its components, and/or those of all previous instances of the process, to:

- make human-based and system-based *cost-informed control-flow decisions*. These decisions would include providing real time calculated values for use as input into branching predicates; to continuously monitor for cost overruns and, when detected, manually or dynamically skip unnecessary or low priority work items, or to cancel work items and/or cases; and to notify administrators when cost thresholds are being approached;
- *allocate work* to resources based on automated decisions about their costs;
- provide human resources and administrators with *cost information* about a process and its component activities, to enable them to make cost-informed decisions about subsequent process executions and process re-engineering;
- support for *cost-informed process variant selections* so that the most appropriate activities are selected and performed based on the current context of the process instance, the calculated costs associated with that context, and archival cost data relevant to the current process execution.

4   Realisation

We have developed an implementation for cost-informed support within the YAWL system environment [3] that fully addresses the success criteria outlined in the previous section. YAWL was chosen as the implementation platform because it is built on an expressive workflow language that provides extensive support for identified workflow and resource patterns. The environment is open-source and offers a service-oriented and vastly extensible architecture, allowing the prototype to be implemented completely independent of the core workflow (enactment) engine.

4.1   The Cost Service

A component service, known as the *Cost Service*, has been realised as a YAWL Custom Service [3]. Satisfying the success criteria using a discrete component provides a solution that remains orthogonal to the main process enactment engine, allowing the approach to be applicable to multiple environments.

The service provides three external REST-based interfaces:

- an *evaluation* interface, which receives notifications from the workflow engine and participating support services at various points in the runtime life-cycle of a process instance, and through which the engine and services may query cost-information, either to request a calculation and have the result returned or to evaluate a cost predicate to a boolean value. Such requests are handled by the *predicate & expression evaluator* sub-component of the service;
- a *log* interface, through which the workflow engine and participating support services may request a complete cost-annotated log of a process instance (or instances) via the *log annotator* sub-component; and
- a *model* interface, which supports the import and export of *cost models*. Such models are stored in the *model handler* sub-component.

Cost Model. A cost model is an XML document that describes all the base cost data and formulae to be associated with a particular process model as defined in [4]. In brief, each cost model consists of three core descriptor sets:

- *Drivers:* Each cost driver defines how cost is associated with one or more process elements (resource, activity, case) together with the relevant cost rate for each element. A cost rate is defined as a data pair of a value and a *per* amount, for example $50 per hour, $70 per tonne, $20 per invocation (fixed) and so on, applied to a process element. An entire process may have any number of drivers defined for it.
- *Functions:* Each function defines an expression for aggregating various cost elements. For example, a function may aggregate a fixed cost, and costs for salaries, insurance and machine hire for all resources involved in an activity.
- *Mappings:* Each mapping provides a method for relating terms used in management accounting to equivalent terms used in a WfMS.

All of the cost model definitions, and the runtime calculations based on them, are managed by the *Cost Service*. Each imported cost model is persisted by the *Cost Service* so that it is always available across future executions of the related process, and across server restarts.

The *Cost Service* is also responsible for the logging of all cost data for all process instances and their activities. The workflow engine and other interested services such as the *Resource Service*, which manages all resourcing allocations, notify the *Cost Service* throughout the life-cycle of each process instance, passing to it the appropriate data so that it can (i) perform the required cost calculations by applying the contextual data (regarding resources, activities, durations, and so on) to the relevant cost model components; and (ii) store all interactions and results in its process logs.

The result of each function call, in addition to being logged by the service, is also returned to the caller for its own decision making and/or logging purposes, as required. Other services may also interface with the *Cost Service* and have cost data logged with activity completion in the engine (via separately defined logging predicates).

Cost-Informed Predicates. The YAWL control-flow predicate syntax has been extended to include cost-informed expressions. A *cost-informed predicate* consists of a conditional expression that may be used to evaluate specified current or historical costs, a comparison operator and a value, and will return a boolean result. Such predicates are associated with control-flow branches at design time, then evaluated at runtime to determine appropriate flow decisions based on the contextual costs of the current process instance.

The syntax of a cost-informed predicate is "**cost(***args***) op numeric_value**", where *args* may comprise:

- zero or one instance of **case()**, which may define within its parentheses a comma separated list of case identifiers and/or case identifier ranges (such as "250-300") and/or any one of the keywords: *max, min* or *average*, and/or any of the keywords *first, last* or *random* followed by an integer (to select a block of cases), and/or either or both of the keywords *from* or *to* followed by a date (to filter on a date range), and/or the keyword *dow* followed by an integer (to filter on day of the week).

– zero or one instance of **task()**, which may define within its parentheses a comma separated list of task identifiers.
– zero or one instance of **resource()**, which may define within its parentheses a comma separated list of resource identifiers.

All arguments act as filters over current and/or historical data. The identifiers and keywords listed in a *case* argument denote that the cost data returned will be limited to those cases. If none are listed inside the parentheses, then all previous and current cases of the current process model will be included. If the entire *case* argument is omitted from the cost predicate, then only the current case is considered.

Some examples of cost expressions:

– **cost()** returns the total cost of the current process instance up to the moment the predicate was evaluated
– **cost(case())** returns the total cost of all past and current instances of the process
– **cost(case(average))** returns the average case cost of all past and current instances of the process
– **cost(from 2014-01-01, dow 2, average)** returns the average case cost of all process instances executed on any Monday since January 1, 2014.
– **cost(case(), task(A), resource(bloggsj))** returns the cost incurred by resource bloggsj performing task A in the current case
– **cost(case(37, 62, 100-145, max), task(A, C, E), resource(bloggsj, 'Senior Manager'))** returns the maximum cost incurred in a single case, of those cases listed, incurred by resource bloggsj or any member of the 'Senior Manager' role while performing the tasks A, C or E

Effect of Cost Predicates at Runtime. In YAWL, each outgoing branch of an OR- (or XOR-) split is associated at design time with a boolean predicate which, if evaluated to true at runtime, determines the branch (or branches, for a multi-OR split) that execution will follow. In this work, the YAWL workflow engine has been extended to accommodate control-flow predicates that include cost-based expressions. When process execution reaches such a predicate, the workflow engine will call the *Cost Service* via its evaluation interface, passing the expression, along with all associated data. The *Cost Service* will use that data, and any archival data as required, to evaluate the expression against the appropriate cost model components, and return the result. The engine will then embed the result into the predicate (replacing the cost-based expression), which it will then continue to evaluate in the usual manner, as needed.

Resource Allocation. The YAWL *Resource Service* provides a pluggable framework for adding new resource *allocation strategies*, which at runtime receive the set of potential resources that may be allocated an activity, and use a defined strategy to select one resource from the set. The standard set of YAWL allocators (e.g. Random Choice, Shortest Queue, Round Robin) has been extended in this work with a number of cost-based strategies, such as *Cheapest Resource, Cheapest to Start, Cheapest Completer* and so on. When a cost-based allocator is enacted at runtime, it will directly query the *Cost Service* via either or both of its evaluation and log interfaces, and may request a calculation based on current and previous case histories (stored within the process logs) for the resources involved, based on its particular allocation strategy. The allocator will then use the result of the query to determine the appropriate resource to whom to allocate the activity, thus fulfilling a *push*-based resource interaction.

A user or administrator interacting with the YAWL worklist, which is also managed by the *Resource Service*, may use a button to invoke a query request to the *Cost Service* for cost data about a selected activity, which will then be displayed in a dialog. The user can then use this information to make ad-hoc cost-informed decisions regarding which activity to choose from their worklist to perform next, thus fulfilling a *pull*-based resource interaction (Fig. 2).

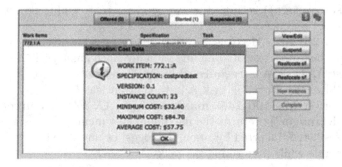

Fig. 2. The YAWL work list extended with cost information (example).

Process Variants. With regards to process variants, the standard YAWL environment includes a service called the *Worklet Service* that allows for the selection of process variants based on the current case context, available data values and an extensible rule set [3]. The rule semantics have been extended to accommodate the same cost-based expression syntax as that used for control-flow predicates. In addition, two comparison functions have been added, called `cheapestVariant()` and `dearestVariant()`, that will apply the associated cost expression against each variant available to a specified activity, and return the variant that is calculated as having the lowest or highest cost of the set, respectively.

When an activity is delegated to the *Worklet Service*, the *Cost Service* is queried for the calculated result of the associated rule from the rule set defined for that activity, which is then used to determine which process variant is the ideal selection based the current, cost-informed context of the case. For example, depending on the cost already expended on a certain case, the service may select a process variant that skips non-essential activities, or depending on the quoted costs from various suppliers of materials, select the variant designed to work directly with the systems of the chosen supplier.

4.2   Cost Service Interaction Architecture

Figure 3 shows the flow of information through the WfMS for each level of cost-informed support.

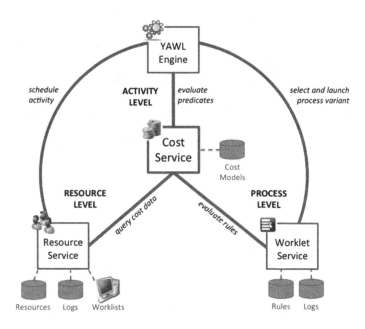

Fig. 3. Cost service interaction architecture.

At the *process* level, the Engine schedules an activity for execution by the *Worklet Service*, which traverses its rule set for the activity, querying the *Cost Service* to evaluate any cost expressions. The *Cost Service* evaluates and returns the result, which the *Worklet Service* uses to select the appropriate process variant for the activity, and launch the variant in the engine.

At the *activity* level, when the Engine encounters a branching construct in the control-flow of a process instance, it queries the *Cost Service* to evaluate any cost expressions in the predicate of each outgoing branch. The engine then uses the results of the predicate evaluations to fire the appropriate branch(es).

At the *resource* level, where the distribution of work takes place, the Engine schedules an activity with the *Resource Service*, which then queries the *Cost Service* for all cost information pertaining to the activity. If the activity is configured for system-based allocation (push pattern), the specified allocation strategy (e.g. Cheapest Resource) is employed using the cost information in its calculations, then the activity is routed to the worklist of the selected resource. If the activity is configured for resource-based allocation (pull pattern), the affected resources' worklists are updated with the retrieved cost information, allowing a resource to select the appropriate activity based on the cost information presented to them.

5   Illustrative Example

The YAWL process model depicted in Fig. 4, adapted from [7], represents a simplified home loan application and approval process that will serve as an example to illustrate the cost-informed support provided by the extensions to the YAWL environment. Tasks that require user interactions are annotated with role-based resource assignments. Most of the tasks in the process are assigned to one specific role, except for three parallel tasks in the *(Re-)Assess Loan Application* sub-process, which can be performed by a resource performing either the Mortgage Underwriter (MU) or Underwriting Assistant (UA) role. There are two tasks that do not require resourcing: *Engage Broker* will be delegated to the Worklet Service for execution of the appropriate variant at runtime, while *Need Mortgage Insurance* task will be executed internally by the YAWL engine using the information provided in the loan application.

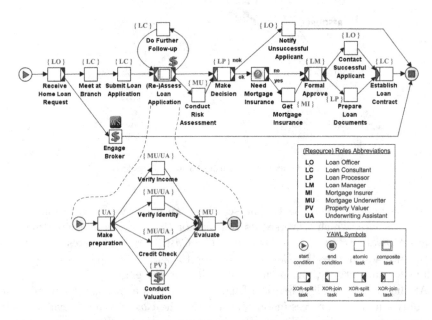

Fig. 4. A home loan process in YAWL (annotated with resource assignments).

The *cost model* for the process is populated based on the assumption that the cost optimisation strategy is to minimise the labour cost for processing a loan case wherever possible. The cost model includes the *cost rates* of resources and activities (modelled as tasks in the process model), of process variants (i.e. models supporting discrete mortgage broker services, managed by the Worklet Service), and of loan application cases (i.e. past and current instances of the process). The cost rates may be specified in different forms as follows:

– Role-based (variable or per hour) cost rate of a resource, e.g. a Loan Officer salary of $40 per hour, or an Underwriting Assistant salary of $50 per hour.
– Fixed cost rate of a resource for a given activity, e.g. to *Conduct [Property] Valuation* a senior Property Valuer charges $300 while a junior Property Valuer charges $250.
– Fixed cost rate of an activity, e.g. *Get Mortgage Insurance* incurs a mortgage insurance processing fee of $50 per loan application (regardless of who conducts the activity).
– Variable cost rate of an activity, e.g. upon *Formal Approval* of a loan application a commission valued at 0.2 % of the loan amount needs to be paid to the Loan Consultant who submitted the application.
– Variable cost rate of a process variant, e.g. a mortgage broker service provider charges a commission valued at 0.5 % of the loan amount.
– Variable cost rate of a case, e.g. a disapproved loan application incurs an additional $100 cost to the case, or an overdue case (of which the duration exceeds the service level agreement) receives a penalty of $150 to the cost of the case.

The cost model is specified as an XML document and uploaded into the Cost Service for processing. The tasks annotated with $ signs in the process model signify those activities where the Cost Service is called to evaluate a course of action during execution. Based on the above cost rates, cost rules are applied to support cost-informed decisions at each of the three levels during process execution.

Process Level. At runtime the Worklet Service handling the *Engage Broker* activity maintains a number of process variants corresponding to various broker services that charge the bank different commissions and/or fees for service provision. Cost-informed rules to guide the selection of a process variant may be based on the cost rate of service providers and/or their performance in previous cases (e.g. case duration). These rules are passed from the Worklet Service to the Cost Service for evaluation. For example, to select the variant that will engage with the broker service that has provided the cheapest average rate over the last 30 cases, the rule would be: *cheapestVariant(case(average, last30))*. Other contextual case data, such as loan amount, the ability to repay, the property location, and planning requirements, can also be used to further filter the result of the business rule.

Activity Level. The XOR-split following completion of the *(Re-)Assess Loan Application* activity is modelled to capture a normal business rule that may also embed a cost-based expression. In normal circumstances, either the *Make*

Decision or the *Do Further Follow-up* activity will be enacted depending on the outcome of the application assessment. These rules are specified as flow predicates of the XOR-split. A flow predicate may also include a cost-informed expression that takes into account the process costs incurred, for example:

- If the (total) processing cost to this point (i.e. loan application assessment) reaches a certain limit (e.g. 0.5 % of the loan amount), then instead of carrying out further follow-up, enact the *Conduct Risk Assessment* activity (expressed as $cost() \geq (loan\_amt * 0.005)$).

Resource level. Multiple resources assigned to perform the same activity may have different cost rates. Consider two typical examples: firstly, both Mortgage Underwriter and Underwriting Assistant can conduct the *Credit Check* activity but are paid a different hourly rate based on their roles; and secondly, even individual resources performing the same role may be paid at different salary levels. It is also likely that individuals performing the same role spend different amounts of time completing certain activities. All past histories of resource completion times for each activity across all process instances (of a particular model) are recorded in a process log, which may be used to define system-based allocation strategies for work distribution that select an available resource from the group of resources that perform the role. Examples are:

- When allocating a certain activity, select the available resource who offers the lowest overall cost rate from the role(s) designated to the activity (the *Cheapest Resource* allocation strategy). This strategy is used in the *Conduct Valuation* activity, where there are individual resources on different pay scales (junior and senior) performing the Property Valuer role.
- When allocating a certain activity, select the available resource who has historically completed the instances of the activity for the least cost, i.e. a function of cost rate and time taken to complete previous instances of the activity (the *Cheapest Completer* strategy). This strategy is used in the *Evaluate* activity, where there are three individual resources performing the Mortgage Underwriter role.

Figure 5 shows screen shots of design-time support for allocation strategy selection in the YAWL designer, while Fig. 6 shows screen shots of the organisational model, the related cost model, and the results of the system-based resource allocation during process execution in the YAWL runtime environment.

6   Related Work

Although cost efficiency is one of the key objectives of a business process management initiative [8,9], there are only a limited number of research studies that examine in detail how cost measures are associated with business processes. The interrelationships between processes, resources and cost are well-known in the

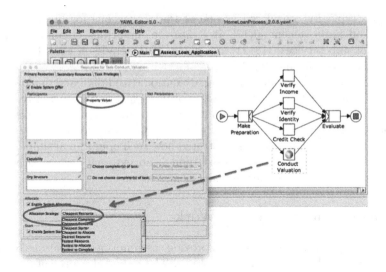

Fig. 5. At design time: the system is configured to allocate the "Conduct Valuation" activity to a resource with Property Valuer role with the cheapest cost.

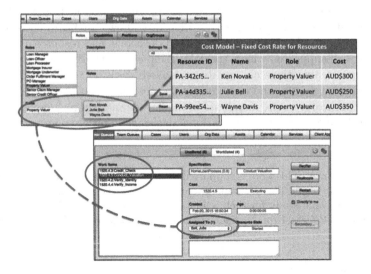

Fig. 6. At runtime: the system selects the resource with the cheapest cost (from the cost model) from all resources performing the "Property Valuer" role (from organisational data) and allocates the work item (instance of activity) "Conduct Valuation" to that resource.

accounting literature [10,11] and several authors have studied the effects of ERP systems on accounting principles [12–14]. Recently, an information model to link the ARIS accounting structure with ARIS process semantics using Event Driven Process Chains (EPC) was proposed in [15] and an extensive set of cost measures to be linked to processes was proposed in [16].

Cost is typically considered as one of the dimensions for business process redesign and improvement [9,17]. It is also an important Quality of Services measure for web services [18–21] and is one of the trade-off considerations for grid and cloud computing [22–24].

Although WfMSs support planning, execution, (re)design and deployment of workflows [25], direct support for cost-informed execution is currently lacking. In our previous work, we presented a generic cost model for cost analysis using event logs [4,26] and a framework to enable cost-informed operational process support [2]. In this paper, we demonstrated how such operational support based on cost considerations can be realised though an external cost service tightly coupled to the YAWL workflow management system.

7   Conclusion and Future Work

The paper proposes an architecture to provide cost-informed operational decision support for workflow management systems. The paper also presents a realisation of such a cost-informed workflow environment within the YAWL workflow management system in the form of a YAWL Cost Service (downloadable from www.yawlfoundation.org). In particular, we demonstrated using a running scenario how cost information can be linked to workflow specification as well as to resource specifications. Furthermore, we illustrated how cost-based decision rules for process variant selections, activity related decisions, and resource assignment decisions are all supported within the YAWL environment. By making use of the Cost Service, organisations can codify cost-based decision rules within a workflow management system and can monitor the cost of operations in real time. In terms of future work, we are interested in the development of techniques that can learn from these cost-informed operational decisions in order to improve process efficiency.

Acknowledgments. This work is supported by an ARC Discovery grant with number DP120101624. We would like to thank all our colleagues who were involved at various stages of this research study for their valuable input. They include Prof. Wil van der Aalst, Prof. Hajo Reijers, Prof. Michael Rosemann, Prof. Zahirul Hoque and Dr. Jochen De Weerdt.

References

1. Chung, L., Nixon, B., Yu, E., Mylopoulos, J.: Non-Functional Requirements in Software Engineering. Kluwer, Dordrecht (2000)
2. Wynn, M.T., Reijers, H.A., Adams, M., Ouyang, C., ter Hofstede, A.H.M., van der Aalst, W.M.P., Rosemann, M., Hoque, Z.: Cost-informed operational process support. In: Ng, W., Storey, V.C., Trujillo, J.C. (eds.) ER 2013. LNCS, vol. 8217, pp. 174–181. Springer, Heidelberg (2013)
3. ter Hofstede, A., van der Aalst, W., Adams, M., Russell, N. (eds.): Modern Business Process Automation: YAWL and Its Support Environment. Springer, Heidelberg (2010)

4. Wynn, M.T., Low, W.Z., Nauta, W.: A framework for cost-aware process management: Generation of accurate and timely management accounting cost reports. In: Conferences in Research and Practice in Information Technology (CRPIT) (2013)
5. van der Aalst, W., Dustdar, S.: Process mining put into context. IEEE Internet Comput. **16**, 82–86 (2012)
6. Russell, N., van der Aalst, W.M.P., ter Hofstede, A.H.M., Edmond, D.: Workflow resource patterns: identification, representation and tool support. In: Pastor, Ó., Falcão e Cunha, J. (eds.) CAiSE 2005. LNCS, vol. 3520, pp. 216–232. Springer, Heidelberg (2005)
7. Wynn, M., De Weerdt, J., ter Hofstede, A., Van Der Aalst, W., Reijers, H., Adams, M., Ouyang, C., Rosemann, M., Low, W.: Cost-aware business process management: a research agenda. In: 24th Australasian Conference on Information Systems (ACIS), pp. 1–11. RMIT University (2013)
8. Kettinger, W., Teng, J., Guha, S.: Business process change: a study of methodologies, techniques, and tools. MIS Q. **21**(1), 55–80 (1997)
9. Reijers, H., Mansar, S.: Best practices in business process redesign: an overview and qualitative evaluation of successful redesign heuristics. Omega **33**(4), 283–306 (2005)
10. Professional Accountants in Business Committee: Evaluating and improving costing in organizations, July 2009
11. Professional Accountants in Business Committee: Evaluating the costing journey: A costing levels continuum maturity model, July 2009
12. Booth, P., Matolcsy, Z., Wieder, B.: The impacts of enterprise resource planning systems on accounting practice-the Australian experience. Aust. Acc. Rev. **10**(22), 4–18 (2000)
13. Grabski, S., Leech, S., Sangster, A.: Management Accounting in Enterprise Resource Planning Systems. CIMA Publishing, Oxford (2009)
14. Hyvönen, T.: Exploring Management Accounting Change in ERP Context. Ph.D. thesis. University of Tampere (2010)
15. vom Brocke, J., Sonnenberg, C., Baumoel, U.: Linking Accounting and Process-Aware Information Systems - Towards a Generalized Information Model for Process-Oriented Accounting. European Conference on Information Systems, pp. 1–13 (2011)
16. Sonnenberg, C., vom Brocke, J.: The missing link between BPM and accounting. Bus. Process Manage. J. **20**, 213–246 (2014)
17. Netjes, M., Reijers, H.A., van der Aalst, W.M.P.: On the formal generation of process redesigns. In: Ardagna, D., Mecella, M., Yang, J. (eds.) Business Process Management Workshops. LNBIP, vol. 17, pp. 224–235. Springer, Heidelberg (2009)
18. Jaeger, M., Rojec-Goldmann, G., Muhl, G.: QoS aggregation for web service composition using workflow patterns. In: Eighth IEEE International Enterprise Distributed Object Computing Conference (EDOC 2004), pp. 149–159. IEEE (2004)
19. Cardoso, J., Sheth, A., Miller, J., Arnold, J., Kochut, K.: Quality of service for workflows and web service processes. Web Semant. Sci. Serv. Agents World Wide Web **1**(3), 281–308 (2004)
20. Eckert, J., Repp, N., Schulte, S., Berbner, R., Steinmetz, R.: An approach for capacity planning for web service workflows. In: 13th Americas Conference on Information Systems (AMCIS 2007). Keystone, Colorado (2007)
21. Mohabbati, B., Gašević, D., Hatala, M., Asadi, M., Bagheri, E., Bošković, M.: A quality aggregation model for service-oriented software product lines based on variability and composition patterns. In: Kappel, G., Maamar, Z., Motahari-Nezhad,

H.R. (eds.) Service Oriented Computing. LNCS, vol. 7084, pp. 436–451. Springer, Heidelberg (2011)

22. Yu, J., Buyya, R.: A budget constrained scheduling of workflow applications on utility grids using genetic algorithms. In: Workshop on Workflows in Support of Large-Scale Science, WORKS 2006, pp. 1–10, June 2006

23. Deelman, E., Singh, G., Livny, M., Berriman, B., Good, J.: The cost of doing science on the cloud: the Montage example. In: Proceedings of the 2008 ACM/IEEE Conference on Supercomputing, SC 2008, vol. 50, pp. 1–12. IEEE Press, Piscataway (2008)

24. Pandey, S., Barker, A., Gupta, K., Buyya, R.: Minimizing execution costs when using globally distributed cloud services. In: 2010 24th IEEE International Conference on Advanced Information Networking and Applications (AINA), pp. 222–229. IEEE (2010)

25. Weske, M.: Business Process Management: Concepts, Languages Architectures. Springer, New York (2007)

26. Wynn, M., Low, W., ter Hofstede, A., Nauta, W.: A framework for cost-aware process management: cost reporting and cost prediction. J. Univ. Comput. Sci. **20**(3), 406–430 (2014)

A Cloud Workflow Modeling Framework Using Extended Proclets

Hua Huang[1,2], Rong Peng[1(✉)], Zaiwen Feng[1], and Min Zhang[2]

[1] State Key Laboratory of Software Engineering, School of Computer,
Wuhan University, Wuhan 430072, China
{rongpeng, zwfeng}@whu.edu.cn,
[2] School of Information Engineering, Jingdezhen Ceramic Institute,
Jingdezhen 333001, China
{jdz_hh, jdz_zm}@qq.com

Abstract. Most existing cloud workflow modeling approaches focus on how to support business process customization for multiple tenants. But there are many factors to be considered in actual cloud workflow modeling, such as how to model interaction between different tenants' business processes while protecting their privacy respectively, and how to facilitate the reuse of common process fragments for different stakeholder roles. To address these issues, this paper proposes a cloud workflow modeling framework using extended proclets. The framework provides a hierarchical management mechanism to isolate the private data of each tenant's business process and adopts two-level-channel transmission protocol to explicitly model interaction between different tenants' business processes. At last, a case study is carried out to illustrate its modeling capability and feasibility.

Keywords: Cloud workflow · Proclets · Interaction between multi-tenant business processes · Hierarchical management mechanism

1 Introduction

Cloud workflow is a workflow management system deployed in cloud computing environment. In cloud workflow, different tenants can design, configure and run their business processes with three-level isolation: data isolation, performance isolation and execution isolation [1]. However, interaction between business process instances of different tenants (e.g. cross-enterprise collaborative business processes) in a cloud workflow system is necessary to be considered. Additionally some common process fragments should be reused for different stakeholder roles. For instance, there exists at least three types of enterprise stakeholder roles (i.e. three types of tenants) in a Cloud Distribution Resource Planning System (denoted as CDRP), i.e. supplier (ceramic product supplier), online distributor (selling suppliers' ceramic products only in online store), physical distributor (selling suppliers' ceramic products in both online store and physical store). Although they share some common processes fragments, e.g., registering and creating online store process, each tenant has its personalized process, e.g., the process of building distribution channel for suppliers and the process of building

© Springer International Publishing Switzerland 2015
J. Bae et al. (Eds.): AP-BPM 2015, LNBIP 219, pp. 19–34, 2015.
DOI: 10.1007/978-3-319-19509-4_2

procurement channel for distributors. In order to complete the product distribution business, interaction between their personalized processes should be achieved. In general, interaction between cross-enterprise collaborative process instances is hidden in customized applications through hard-coded business process definition languages, which contributes to the difficulties for updating and maintaining the corresponding system. Therefore, it is necessary to explicitly model interaction between different stakeholder roles in CDRP. Unfortunately, another challenge emerges when modeling interaction between multi-tenant business processes. That is how to isolate and protect each tenant's business processes related private data [2].

To solve these problems, we adopt hierarchical management ideology to extend the proclets framework [5, 6] for cloud workflow modeling. In summary, we make the following contributions in this paper:

- Propose a novel cloud workflow modeling framework based on extended proclets, which adopts two-level-channel transmission protocol to explicitly model interaction between different tenants' business processes. Meantime, it provides a hierarchical management mechanism to protect their private data.
- A case study is conducted within a CDRP System to demonstrate the feasibility of the proposed approach.

The remainder of this paper is organized as follows. Section 2 discusses related work. Section 3 proposes a cloud workflow modeling framework using extended proclets. In Sect. 4, we illustrate our approach by modeling three types of tenants' (suppliers, online distributors, physical distributors) business processes in CDRP. Section 5 concludes this paper.

2 Related Work

Recently, cloud-based business process modeling approaches are emerging. For instance, Mietzner [7] presents a customizable process model by individualizing a given process template based on tenant's requirements, and then converting it into a BPEL process model. In [8], BPEL is extended into VxBPEL, which enables business processes to be individualized based on variable points. And Ralph also adopts variable points to customize BPEL process model [7, 9, 10]. However, most cloud-based business modeling methods focus on multi-tenant business process customization without considering interactions between different tenant business processes.

Meanwhile, the tools and languages for modeling cloud-based business processes are also to extend the general modeling notation, e.g. Petri net, BPMN, EPC, UML AD, BPEL, etc. These approaches focus on isolated case-based processes without considering the interaction between multiple business processes. In recent years, the object-oriented and artifact-centric modeling approaches provide a solution combining process-centric and data-centric process modeling approaches [3, 4]. The object-oriented modeling approaches adopt objects to denote the information entities that capture business goals [3]. Compared with traditional business process modeling approaches, they focus on modeling business contexture and behavior, rather than activity sequences. With respect to the artifact-centric modeling approaches, the business process is defined by four components: business artifact, life-cycle, business rules and services [4]. These approaches use business artifacts to combine data-flow and control-flow in a holistic manner as the basic building blocks. Thus, it is easier to implement the migration of business processes. Although the advantages of the object-oriented and artifact-centric modeling approaches have been evaluated both in academic research and industrial applications, they have not yet been extended to cloud workflow modeling. In addition, there is another business process modeling approach based on proclets framework [5, 6]. A proclet is light-weight workflow process, which decomposes a big and complex process into several interacting sub-processes. The proclets framework addresses the issues of communication and collaboration of internal sub-processes. Moreover, it is possible to model complex workflows in a more natural manner by proclets. It means that the proclets framework is helpful for solving the problems mentioned in Sect. 1, e.g., interaction between multi-tenant business processes. In order to use proclets to model cloud workflow, it is necessary to extend proclets framework based on hierarchical management ideology.

3 Cloud Workflow Modeling Framework Using Extended Proclets

The proclets framework is composed of four main components: proclet, channel, naming service, and actor. The original framework of proclets [6] is shown in Fig. 1.

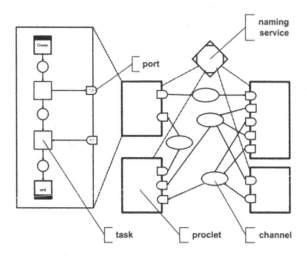

Fig. 1. The original framework of proclets

In Fig. 1, a proclet can be seen as a light-weight workflow or a process object with life-cycle. It can interact with other proclets via communication channels, which are the medium to transmit messages from one proclet to another. A task (or transition) can send or receive messages via ports. In the original framework, all proclet instances are monitored and managed by a naming service. The detail of each component can be found in [6]. To satisfy new requirements for cloud workflow modeling, we adopt hierarchical management method to extend channel, naming service, knowledge base and proclet class except for actor, which is packaged into proclet class as process related data (namely role).

3.1 Extending Channel

Channels are used to link multiple proclets and transmit messages between them. To achieve transmitting messages between different tenant-level proclets (Definition in Sect. 3.4) in cloud workflow, we extend channel into two categories: tenant-level channel and tenant-inner-level channel. Tenant-level channel (depicted as a double-solid-line ellipse) is used to link tenant-level proclets for transmitting messages between them. Tenant-inner-level channel is similar to the channel in Fig. 1, but it includes three types: tenant-inner-level input channel (denoted as *Inner Channel for Input*), tenant-inner-level output channel (denoted as *Inner Channel for Output*) and tenant-inner-level communication channel (denoted as *Inner Channel*). Tenant-inner-level input channel (depicted as a single point-line ellipse) is used to link tenant-level proclets and tenant-inner-level proclets (Definition in Sect. 3.4) for transmitting messages from tenant-level proclets to tenant-inner-level proclets, while tenant-inner-level output channel (depicted as a single-dotted-line ellipse) is used to link tenant-inner-level proclets and tenant-level proclets for transmitting messages from tenant-inner-level proclets to tenant-level proclets. Tenant-inner-level communication channel (depicted as a single-solid-line ellipse) is used to link tenant-inner-level proclets for transmit messages between them.

3.2 Extending Naming Service

In Fig. 1, all interaction between proclets is based on proclet identifiers (*proc_id*) and class identifiers (*class_id*). These identifiers provide the handles to route communication information. In general, the sending proclet does not know the *proc_id* of all receiving proclets. Thus, the original proclets framework [6] adopts naming service to keep track of all proclets. The naming service provides four basic functions: registering, updating, querying, and unregistering proclets' related information, such as *proc_id, class name, creator, creating time, owner* and *key attributes*. In the extended proclets framework, the naming services can be grouped into two categories: global naming service (depicted as a double-line diamond) and inner naming service (depicted as a single-line diamond). In a cloud workflow system, there is only one global naming service, which is used to register, query, update, and unregister all tenant-level proclets. The inner naming service is defined in a tenant-level proclet. For any tenant, there exists only one tenant-level proclet, thus there is only one inner naming service to manage its tenant-inner-level proclets.

3.3   Extending Knowledge Base

A proclet uses knowledge in its knowledge base to make routing decisions. In order to support interactive communication between tenant-level proclets, we extend knowledge base into two categories: tenant-level knowledge base and tenant-inner-level knowledge base. Tenant-level knowledge base is defined in a tenant-level proclet. It mainly consists of a communication log table, which is used to keep all records of communication information related to the tenant-level proclet, including communication time, channel name, sender, *proc_id* of sender's tenant-level proclet, receivers, *proc_ids* of receivers' tenant-level proclets, action, scope (Public or Private), etc. While tenant-inner-level knowledge base is defined in a tenant-inner-level proclet. It also mainly contains a communication log table, in which the stored information is similar to the tenant-level knowledge base except for *proc_id* of sender's tenant-level proclet and *proc_ids* of receivers' tenant-level proclets.

3.4   Extending Proclet Class

In Fig. 1, the life-cycle of a proclet instance and all its ports is described via a proclet class (denoted as *PC*), which can be compared to an ordinary workflow process definition or workflow type. In the original proclets framework, the proclet class describes the execution order of all tasks using a graphical language based on Petri nets. In this paper, we adopt the artifact-centric modeling approach to package the process related data (e.g., the executor and service resources needed for implementing tasks of business processes, etc.) and the life-cycle of a proclet instance into a proclet class. The activity sequences involved in the life-cycle are represented by Petri nets. To support hierarchical management, proclet class is extended into two categories: tenant-level proclet (denoted as *Tenant_PC*) and tenant-inner-level proclet.

   Tenant_PC is a virtual process object created for each tenant. It is used to monitor and manage all tenant-inner-level proclets. A tenant-level proclet class has a unique identity (*class_id*). Meanwhile, each tenant corresponds to only one tenant-level proclet instance with unique identity (*proc_id*). The *proc_id* is consistent with tenant account (denoted as *tenant ID*). A tenant-level proclet class is composed of four parts: tenant-level knowledge base, tenant process related data set, life-cycle, and inner naming service. The tenant process related data set contains a set of tenant related attributes (*tenant ID, tenant name, tenant type, tenant profile*), a set of tenant-inner personnel roles and a set of service resources for implementing tenant business processes. The class structure of *Tenant_PC* is shown in Fig. 2.

   In Fig. 2, the life-cycle of *Tenant_PC* is composed of four transitions (*Create, Input Processing, Output Processing* and *End*), three places (*P1,P2,P3*) and four ports (*Outer Input, Outer Output, Inner Output, Inner Input*). The transition *Create* is activated to generate its instance when a tenant is registering. The transition *End* is executed to destroy its instance when a tenant is unregistering. The transition *Input Processing* is linked by two ports: *Outer Input* and *Inner Output*. The port *Outer Input* is used for receiving communication messages from other *Tenant_PC* instances through the tenant-level channel, then the communication messages are submitted to the transition

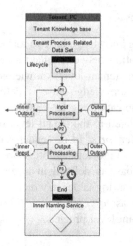

Fig. 2. The class structure of *Tenant_PC*

Input Processing for handling operation, e.g., checking message, decrypting ciphertex, sending classification and saving log (storing the routing information into the log table of tenant-level knowledge base), etc. The port *Inner Output* is used to send communicate messages handled by the transition *Input Processing* to the receiver (a tenant-inner-level proclet instance) through the tenant-inner-level input channel. The transition *Output Processing* is also linked by two ports: *Inner Input* and *Outer Output*. The port *Inner Input* is used for receiving communication messages from tenant-inner-level proclet instances through the tenant-inner-level output channel, then the communication messages are submitted to the transition *Output Processing* for handling operation, e.g., coding message, encrypting related private data, sending classification and saving log, etc. The port *Outer Output* is used to send communicate messages handled by the transition *Output Processing* to other *Tenant_PC* instances through the tenant-level channel.

Tenant-inner-level proclet is an actual process object, which is used to encapsulate each tenant's business process and its related business data with a certain size granularity. It is divided into two categories: shared tenant-inner-level proclet (denoted as *Shared_PC*) and general tenant-inner-level proclet (denoted as *Inner_PC*).

Shared_PC is used to encapsulate common process objects and related metadata for the same type tenants. According to whether *Shared_PC* can be customized, it can be extended into two categories: fixed shared tenant-inner-level proclet (denoted as *Fixed_Shared_PC*) and customizable shared tenant-inner-level proclet (denoted as *Customizable_Shared_PC*). *Fixed_Shared_PCs*, which are common business process objects pre-designed and stored in process repository, can be reused without any modification. Whereas they cannot be customized. As shown in Fig. 3, a *Fixed_Shared_PC* class is composed of three main components: tenant-inner-level knowledge base, shared process related data set and life-cycle. The life-cycle of a *Fixed_Shared_PC* class is mainly composed of many transitions and places. A transition is linked by ports (Input Port or Output Port). The Input Port is used for receiving communication messages transmitted through tenant-inner-level channels, and the Output Port is used for sending communication messages from the linked transition. The shared process related data set mainly contains a set of public business attributes (e.g., *business ID, business name, business type, business description*, etc.) and a set of relations between transitions and resources (including roles and services used for executing business activities), etc. *Customizable_Shared_PCs* are also common business process objects pre-designed in cloud workflow system or defined by tenants. They can be reused by making modification (e.g., adding or deleting transitions, setting the input and output ports, modifying business metadata, etc.). As opposed to *Fixed_Shared_PCs*, they can be customized according to tenants' requirements. As shown in Fig. 3, a *Customizable_Shared_PC* class is composed of three components: tenant-inner-level knowledge base, customizable process related data set and life-cycle. Although the life-cycle and customizable process related data set of *Customizable_Shared_PC* is similar to the definition of *Fixed_Shared_PC*, the former can be customized and the latter cannot.

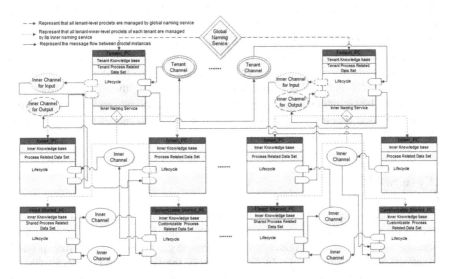

Fig. 3. The cloud workflow modeling framework based on extended proclets

Inner_PC is defined by tenants. During the modeling process, according to multiple business goals or business artifacts, a big and complex business process can be decomposed into multiple sub-processes that can be easily modeled and implemented. These sub-processes can be modeled as *Inner_PCs*. As shown in Fig. 3, a *Inner_PC* class is composed of three components: tenant-inner-level knowledge base, process related data set and life-cycle. The life-cycle and process related data set of *Inner_PC* is similar to the definition of *Customizable_Shared_PC*.

According to the above description for extending proclets, the cloud workflow modeling framework based on extended proclets is proposed as shown in Fig. 3.

In Fig. 3, there exists only one global naming service, which is used for monitoring and managing all tenant-level proclet instances. For each tenant, all tenant-inner-level proclet instances belonging to it are monitored and managed by its inner naming service.

Interaction between tenant-inner-level proclet instances belonging to the same tenant-level proclet instance is easily achieved by tenant-inner-level communication channels. However, it is difficult to achieve interaction between tenant-inner-level proclet instances belonging to different tenant-level proclet instances. For example, in Fig. 3, a tenant-inner-level proclet instance (assumed as *PC1*) owned by a tenant-level proclet instance (assumed as *Tenant_PC1*) sends a communication message (assumed as *MSG*) to another tenant-inner-level proclet instance (assumed as *PC2*) owned by another tenant-level proclet instance (assumed as *Tenant_PC2*). Firstly, *MSG* is created by a transition (assumed as *T1*) in *PC1*, then *MSG* is sent by the Output Port linked to *T1*, and transmitted to the Inner Input Port of the *Tenant_PC1* through the tenant-inner-level output channel. *MSG* is submitted to the transition *Output Processing* of *Tenant_PC1* for security handling (e.g., security coding, data encryption, etc.). After having been handled, *MSG* is sent by the Outer Output Port of *Tenant_PC1*, and transmitted to the Outer Output Port of *Tenant_PC2* through the tenant-level channel. Then, *MSG* is submitted to the transition *Input Processing* of *Tenant_PC2* for security handling (e.g., security authentication, cipher text decryption, etc.). After having been handled, *MSG* is sent by the Inner Output Port of *Tenant_PC2*, and transmitted to the Output Port linked to the receiver transition (assumed as *T2*) in *PC2* through the tenant-inner-level input channel. Finally, *T2* receives *MSG* successfully and be fired. It means that one interaction between *PC1* and *PC2* is completed without disclosing their private data.

In order to better understand the proposed cloud workflow modeling framework based on extended proclets, a case using this framework to model tenants' business process in CDRP is presented as below.

4   Case Study

The CDRP mentioned in Sect. 1 is designed to enhance the communication and interaction between supplier, online distributor, and physical distributor and to implement online ceramic product sale business, e.g., publishing ceramic product information, managing order, etc. In order to apply the services in CDRP, all tenants need to register and create online stores. For suppliers, when having registered and created their online stores, they are able to build distribution channels and online supply ceramic products. For online distributors or physical distributors, they also need to build purchasing channels to online purchase and sell ceramic products after having registered and created their online stores. It is clear that there exists interactions between the distribution channel processes built by suppliers and the purchasing channel processes built by distributors. In addition, there exists interactions between suppliers' online supplying ceramic products processes and distributors' online purchasing and selling ceramic products processes. Meanwhile, for online distributors and physical distributors, there are some differences in purchasing and selling ceramic products process. The online distributors make online selling ceramic products without stock, so they can directly download ceramic products authorized by their suppliers to their shops for sale. When a buyer creates an order in a online store, the corresponding purchasing order will be generated to the corresponding supplier. With respect to physical distributors, they can download ceramic products authorized by their suppliers after having purchased them. Then, they can sell these ceramic products online or offline in their online or physical stores.

4.1 Description of Modeling Each Process Object

According to the aforementioned description in CDRP, three types of tenants (suppliers, online distributors and physical distributors) have their own business processes when they are using the services provided by CDRP. Among these processes, some sub-process or process fragments are identical or similar. To reuse them and simplify business process modeling for each tenant, tenant's business process are decomposed into multiple sub-processes based on sub-targets within a certain granularity; then, package each sub-process and related process data into a process object defined by a proclet. Thus, the supplier's business process can be decomposed into three sub-process objects: registering and creating online store, building distribution channels, and supplying ceramic products. The online distributor's business process can be decomposed into four sub-processes objects: registering and creating online store, building the purchasing channels, purchasing ceramic products and online selling ceramic products. It is the same as the physical distributor's business process except for online or offline selling ceramic products. Obviously, the common sub-process of registering and creating online store can be used by three types of tenants. Therefore, it is modeled as a *Fixed_Shared_PC* (denoted as *Create_Shop_Fixed_Shared_PC*). The sub-process of building the purchasing channels, in which there are some differences between online distributors and physical distributors, is modeled as a *Customizable_Shared_PC* (denoted as *Building_Procurement_Channel_Customizable_Shared_PC*). This is the same as the sub-process of purchasing ceramic products (denoted as *Product_Purchase Customizable_Shared_PC*). Other sub-processes will be modeled as *Inner_PCs*, i.e. *Building_Distribution_Channel_Inner_PC*, *Product_Supply_Inner_PC*, *Online_Sale_Inner_PC* and *Online_and_Offline_Sale_Inner_PC*. For better understanding how to define three types of *Inner_PCs*, we select three *Inner_PCs*: *Create_Shop_Fixed_Shared_PC*, *Building_Procurement_Channel_Customizable_Shared_PC* and *Online_Sale Inner_PC* as examples to describe their model structure.

(1) *Create_Shop_Fixed_Shared_PC*
The sub-process object of registering and creating online store (denoted as *Create_Shop_Fixed_Shared_PC*) is pre-designed as a *Fixed_Shared_PC* in CDRP. It can be reused for three type of tenants when they are modeling their business processes. Meanwhile, it can be instantiated at runtime and monitored by the tenant-level proclet. Its model diagram is shown in Fig. 4.

In Fig. 4, the message places are marked up as P1 to P5. The transition *Create* is used to create proclet instance, and the transition *End* is used to destroy proclet instance. There are other transitions marked up as T1 to T4, the details of them are described as below.

T1: Register information.
T2: User login.
T3: Create online store.
T4: Decorate online store.

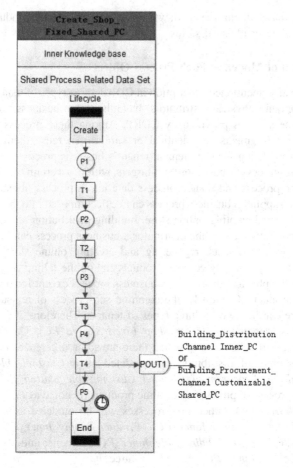

Fig. 4. Model diagram of *Create_Shop_Fixed_Shared_PC*

POUT1: An output port, one end of which is connected to T4, and the other end is pointed to *Building_Distribution_Channel_Inner_PC* or *Building_Procurement _Channel_Customizable_Shared_PC*. This means that after having decorated online store (or T4 is executed), for suppliers, T4 will send a communication message to activate *Building_Distribution_Channel_Inner_ PC* though POUT1; but for distributors, T4 will send a communication message to activate *Building_Procurement_Channel_Customizable_ Shared_PC*.

(2) *Building_Procurement_Channel_Customizable_Shared_PC*
The sub-process object of building procurement channels (denoted as *Building_Pro-curement_Channel_Customizable_Shared_PC*) is pre-desinged as a *Customizable_Shared_PC* in CDRP. It is used to apply for distribution eligibility and download products for distributors. For online distributors, they can directly download the

ceramic products authorized by their suppliers, yet physical distributors can download the ceramic products authorized by their suppliers only after having purchased them. Therefore, the communication message of finishing order is necessary to be considered. Its model diagram is shown in Fig. 5.

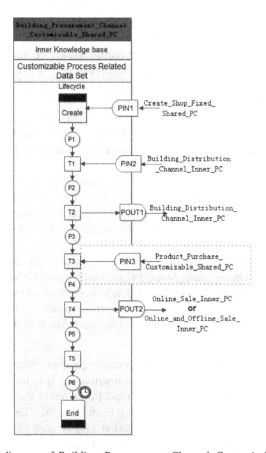

Fig. 5. Model diagram of *Building_Procurement_Channel_Customizable_Shared_PC*

In Fig. 5, the message places are marked up from P1 to P6. The transition *Create* is used to create the proclet instance, the transition *End* is used to destroy the proclet instance. There are other transitions marked up from T1 to T5. The details are described as below.

T1: Find suppliers.

T2: Apply for distribution eligibility.

T3: For online distributors, T3 represents downloading ceramic products authorized by their suppliers; for physical distributors, T3 represents downloading ceramic products purchased by them.

T4: Show ceramic products for online sale.

T5: Manage suppliers' information.

PIN1: An input port, one end of which is connected to the transition *Create*, and the other end is directed by *Create_Shop_Fixed_Shared_PC*. This means that the transition *Create* will be activated after having received the activating message from *Create_Shop_Fixed_Shared_PC* through PIN1.

PIN2: An input port, one end of which is connected to T1, and the other end is directed by *Building_Distribution_Channel_Inner_PC*, which indicates that T1 will be activated after having received the message of recruiting distributors from *Building_Distribution_Channel_Inner_PC* through PIN2.

PIN3: An input port, one end of which is connected to T3, and the other end is directed by *Product_Purchase_Customizable_Shared_PC*. PIN3 is included in a dashed box, which means that PIN3 is defined only in *Building_Procurement_ Channel_Customizable_Shared_PC* for physical distributors. Having received the message of finishing procurement transactions from *Book_Purchase_Customizable _Shared_PC* through PIN3, T3 will be activated.

POUT1: An output port, one end of which is connected to T2, and the other end is pointing to *Building_Distribution_Channel_Inner_PC*. This means that after having submit the application for distribution eligibility, T2 will send the reviewing message to *Building_Distribution_Channel_Inner_PC* through POUT1.

POUT2: An output port, one end of which is connected to T4, and the other end is pointing to *Online_Sale_Inner_PC* or *Online_and_Offline_Sale_Inner_PC*. This means that after T4 is executed, it will send the communication message to *Online_Sale_Inner_PC* for online distributors or to *Online_and_Offline_Sale_Inner_PC* for physical distritutors through POUT2.

(3) *Online_Sale_Inner_PC*

The sub-process object of online selling ceramic products (denoted *Online_Sale_Inner_PC*) is an *Inner_PC* defined by online distributors. It is used to sell suppliers' ceramic products for online distributors. Its model diagram is shown in Fig. 6.

In Fig. 6, the message places are marked up from P1 to P6. The transition *Create* is used to create the proclet instance, the transition *End* is used to destroy the proclet instance. There are other transitions marked up from T1 to T5. Their details are described as below.

T1: Buyer purchases ceramic products and create orders online.
T2: Buyer pays to online distributors.
T3: Online distributors confirms receivables and notify suppliers for delivery.
T4: Buyer confirms the receipt and finishes the transaction.
T5: Manage sale orders.

PIN1: An input port, one end of which is connected to the transition *Create*, and the other end is directed by *Building_Procurement_Channel_Customizable_Shared_PC*. This means that the transition *Create* will be activated after having received the activating message from it through PIN1.

POUT1: An output port, its one end is connected to T1, and the other end is pointed to *Product_Purchase_Customizable_Shared_PC*. It indicates that after buyer

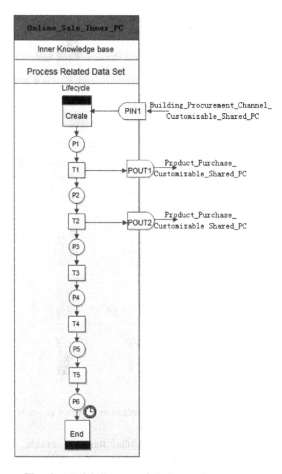

Fig. 6. Model diagram of *Online_Sale_Inner_PC*

created orders, T1 will send the notification message to *Product_Purchase_ Customizable _Shared_PC* through POUT1.

POUT2: An output port, one end of which is connected to T2, another is pointing to *Product_Purchase_Customizable_Shared_PC*. This means that after buyers having pay to the online distributors, T2 will send the notification message to *Product_Purchase_Customizable_Shared_PC* through POUT2.

4.2   Building Interaction Between Proclets

In this section, we model interaction between proclets defined in CDRP. The cloud workflow model built for CDRP is shown in Fig. 7 (to simplify the drawing, the life-cycle of each proclet is omitted, while the input and output ports are depicted).

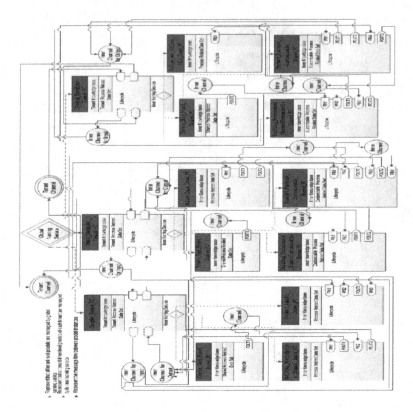

Fig. 7. Model diagram of interaction between proclets

In Fig. 7, the model consists of one global naming service, three tenant-level proclets (respectively denoted as *Suppler_Tenant_PC*, *Online_Distributor_Tenant_PC*, and *Physical_Distributor_Tenant_PC*), some *Inner_PCs* and channels used for interactions between proclets. The working principle of the model involves three aspects: process object reuse, hierarchical management and two-level-channel transmission, the details are given as below.

Process Object Reuse. In CDRP, *Create_Shop_Fixed_Shared_PC* is a *Fixed_Shared_PC*, which can be reused without conducting any modification for three types of tenants. In addtion, *Building_Procurement_Channel_Customizable_Shared_PC* and *Product_Purchase_Customizable_Shared_PC* are *Customizable_Shared_PC*. They can be reused by customizing the transitions and ports of their life-cycle, and configuring the process related data set according to distributors' business requirements.

Hierarchical Management. In Fig. 7, there exists two-level management: tenant-level management and tenant-inner-level management, which can isolate the related private data of tenant business process. In tenant-level management, all tenant-level proclets (i.e. *Suppler_Tenant_PC*, *Online_Distributor_Tenant_PC* and *Phyical Distributor_Tenant_PC*) are monitored and managed by only one global naming service. In tenant-inner-level management, all *Inner_PCs* belonging to a *Tenant_PC* are monitored and managed by the inner naming service defined in the *Tenant_PC*. For example, in Fig. 7, suppliers' process is composed of three *Inner_PC*s: *Create_Shop_Fixed_Shared_PC*, *Building_Distribution_Channel_Inner_PC* and *Product_Supply_Inner_PC*. They are monitored and managed by the inner naming service defined in *Suppler_Tenant_PC*. This is the same as distributor's process.

Two-level-channel Transmission. In Fig. 7, there exist two types of transmissions: transmission through tenant-inner-level channels and transmission through both tenant-level channels and tenant-inner-level channels. Interaction between tenant-inner-level proclets belonging to the same tenant-level proclet is achieved by the first transmission. For example, in supplier tenant's business process, *Create_Shop_Fixed_Shared_PC* sends an activating message to *Building_Distribution_Channel_Inner_PC* through the tenant-inner-level channels. To achieve the interaction between tenant-inner-level proclets belonging to different tenant-level proclets, the second transmission is adopted. For instance, in Fig. 7, the process of T3 in suppliers' *Building_Distribution_Channel_Inner_PC* sending a message of recruiting distributors to T1 in online distributors' *Building_Procurement_Channel_Customizable_Shared_PC* is shown as follow. Firstly, the message is sent by the POUT1 connected to T3, and transmitted to the transition *Output Processing* of *Suppler Tenant_PC* for security handling (e.g., security coding, data encryption, etc.) through the tenant-inner-level output channel. Then, the message is transmitted to the transition *Input Processing* of *Online_Distributor Distrituctor Tenant_PC* for security handling (e.g., security authentication, ciphertext decryption, etc.) through the tenant-level channel. Finally, the message is transmitted to PIN2 through the tenant-inner-level input channel, and T1 is fired after receiving the message. Now, one interaction between *Building_Distribution_Channel_Inner_PC* and *Building_Procurement_Channel_Customizable_Shared_PC* is successfully achieved without disclosing their process related private data.

The above proposed case exeplifies that our modeling framework can explicitly model the interactions between multi-tenant business processes while protecting their private data and reusing common process fragments.

5   Conclusion

In this paper, we extend the proclets approach based on hierarchical management ideology and propose a cloud workflow modeling framework using extended proclets. This framework is successfully applied for designing and implementing the CDRP, which can be accessed by http://www.ccmall.cn. Compared with the existing cloud-based workflow modeling approaches, our approach has the following characteristics:

(1) In most existing cloud workflow modeling approaches, a tenant's business process is modeled into a big and complex process model, which has too many tasks and too complex control logic. Therefore, it is difficult to execute and monitor. Meanwhile it may increase the error probability of process executing. In our framework, each tenant's business process is decomposed into several sub-process objects according to its sub-goals or key artifacts, then these sub-processes are abstracted and packaged into tenant-inner-level proclets. During the modeling process, common sub-processes

belonging to multiple tenants are modeled as *Fixed_Shared_PCs* or *Customizable_Shared_PCs*. It can greatly improve the efficiency of business process modeling for each tenant in cloud workflow system.

(2) The existing cloud workflow modeling frameworks do not support the isolation and protection of the private data related to tenant business processes. In our framework, it can isolate and protect the related private data of tenant business processes by hierarchical modeling and management.

(3) The existing cloud workflow modeling approaches do not model interaction between multi-tenant business processes. In this paper, we adopt two-level-channel transmission to explicitly model interactions between multi-tenant business processes.

In summary, the proposed modeling framework using extended proclets is feasible to model interactions between multi-tenant business processes without disclosing their private data. It also provides a model for designing cloud workflow engine.

Acknowledgments. This work is supported by the National Natural Science Foundation of China under Grant No. 61170026, 61100017, the National Science and Technology Ministry of China under Grant Nos. 2012BAH25F02, 2013BAF02B01 and the Fundamental Research Funds for the Central Universities of China under Grant No. 2012211020203, 2042014kf0237.

References

1. Chen Kang, S., Wei-min, Z.: Cloud computing: system instances and current research. J. Softw. **20**(5), 1337–1348 (2009)
2. Deng-Guo, F., Min, Z., Yan, Z., Zhen., X.: Study on cloud computing security. J. Softw. **22**(1), 71–83 (2011)
3. Redding, G., Dumas, M., et al.: A flexible object-centric approach for business process modeling. SOCA **4**(3), 191–201 (2010)
4. Cohn, D., Hull, R.: Business artifacts: a data-centric approach to modeling business operations and processes. IEEE Data Eng. Bull. **32**(3), 3–9 (2009)
5. van der Aalst, W.M.P., Mans, R.S., Russell, N.C.: Workflow support using proclets: divide, interact, and conquer. In: Proceedings of IEEE Data Engineering Bullet, pp.16–22 (2009)
6. van der Aalst, W.M.P., Barthelmess, P., Ellis, C.A., Wainer, J.: Proclets: a framework for lightweight interacting workflow processes. Int. J. Coop. Inf. Syst. **10**(4), 443–481 (2001)
7. Mietzner, R., Leymann, F.: Generation of BPEL customization processes for SaaS applications from variability descriptors. In: Proceedings of the 2008 International Conferences on Services Computing (SCC 2008), pp. 359–366 (2008)
8. Michiel, K., Chang-ai, S., Marco, S., Paris, A.: VxBPEL: supporting variability for web services in BPEL. Inf. Softw. Technol. **51**, 258–269 (2009)
9. Ralph, M.: Using Variability Descriptors to Describe Customizable SaaS Application Templates, pp. 1–27. Institute of Architecture of Application Systems, 18 January 2008
10. Ralph, M., Andreas, M., Frank, L., Klaus, P.: Variability modeling to support customization and deployment of multi-tenant-aware software as a service applications. In: Proceeding of the 2009 ICSE Workshop on Principles of Engineering Service Oriented Systems (PESOS 2009), pp.18–25 (2009)

A Gateway-Centered Workflow Rollback-Points Ancestry Model for Sustainable Workflow Enactments

Minjae Park[1], Hyun Ahn[2], Haksung Kim[2], and Kwanghoon Pio Kim[3]([⊠])

[1] Engineering Innovation Department, R&D Center, BISTel, Inc.,
BISTel Tower, 128 Baumoe-ro Seocho-gu Seoul, Seoul 137-891, South Korea
mjpark@bistel-inc.com
http://www.bistel.co.kr
[2] Department of Taxes and Accounting, Dongnam Health University,
74 Chunchun-ro Jangan-gu, Suwon-si, Kyonggi-do 440-714, South Korea
amang@dongnam.ac.kr, hahn@kgu.ac.kr
http://www.dongnam.ac.kr
[3] Collaboration Technology Research Lab, Department of Computer Science,
Kyonggi University, 154-42 Kwangkyosan-ro Youngtong-ku,
Suwon-si, Gyeonggi-do 443-760, South Korea
kwang@kgu.ac.kr
http://ctrl.kyonggi.ac.kr

Abstract. In enacting a workflow process that is made up of stepwise activities and their temporal orderings, it is very important to control and trace each instance's execution as well as to keep it sustainable. In particular, the sustainability implies that the underlying system should be able to not only provide an error-detection functionality on its running exceptions but also to furnish a very autonomous recovery mechanism to deal with the detected exceptional and risky situations. As an impeccable technique for realizing the sustainability on workflow process enactment services, this paper tries to formalize a workflow risk-recovery concept to be used in implementing autonomous recovery mechanisms of workflow enactment systems, which is named as *gateway-centered rollback-points ancestry model.* Conclusively, we believe that the proposed model ought be one of those impeccable trials and pioneering contributions to improve and advance the sustainability in workflow process enactment services.

Keywords: Information control net · Workflow process · Risk dependency · Sustainability · Autonomous workflow recovery · Rollback-points ancestry model

1 Introduction

In this paper, we focus on the sustainability issue [8, 11, 12] in enacting workflow processes by a workflow management system. In order to guarantee safeness on

This research is mainly supported by the Contents Convergence Software Research Center funded by the GRRC Foundation of Gyeonggi Province, South Korea.

© Springer International Publishing Switzerland 2015
J. Bae et al. (Eds.): AP-BPM 2015, LNBIP 219, pp. 35–42, 2015.
DOI: 10.1007/978-3-319-19509-4_3

workflow processes synchronization that implies to keep consistency between the run-time information on the system and the physical status of a corresponding business process on the real-world, it is very important for the system to be supported by an autonomous error-detection and self-recovery mechanism for resolving the exceptional and risky situations. A little more specifically speaking, the paper conceives a novel concept of rollback-points ancestries to be used for the system to automatically recover and resume the error-involved workflow instances from the exceptional and risky situations. The proposed concept implies that the self-recovery mechanism is able to decide a proper point of rollback operations, which is pointing to the activity (either gateway-activity or task-activity) starting a rollback among the previous performed activities in the corresponding workflow instance, according to the guidance of a rollback-points ancestry model.

In principle, there are two types of rollback-points ancestry models, such as gateway-centered rollback-points model and task-centered rollback-points ancestry model. A rollback-points ancestry model basically specifies a series of rollback sequences as workflow instance recovery information, which can be used for implementing an autonomous self-recovery mechanism to resume the error-involved workflow instances from the exceptional and risky situations. We formalize the gateway-centered rollback-points ancestry model and its usage, in particular. Note that a workflow process model describes a temporal precedence (control flow) of gateway activities and task activities, and its related events, such as initiating, terminating, timing, and so on. So, the rollback-point of a workflow process model ought to be either a gateway activity or a task activity. That is, once the system detects an error situation, and then it has to make a decision for choosing a rollback-point where the system applies a series of forward-rollback operations to recover the corresponding workflow instance. In general, we can use three types of gateway activities [3], such as alternative-gateway, conjunctive-gateway, and iterative-gateway, in defining a workflow process model, and so the gateway-centered rollback-points ancestry model proposed in the paper specifies a temporal sequence of gateway-activities that can be helped for the system to choose a reasonable rollback-point, efficiently as well as autonomously. In this paper, we expatiate on the details of gateway-centered workflow rollback-points ancestry model with an operational example.

2 Conceptual Backgrounds and Related Work

We know that the system's self-recoverable ability depends on its rollback mechanism to be applied to those running workflow instances. The rollback mechanism used to determine a rollback-point by indicating an activity out of the previously completed activities on the underlying workflow instance. However, because the system does not know from where the detected exceptions come, it has to perform the rollback operations of all the completed activities to abort the running workflow instance. At this moment, we also know that the system ought to be much more efficient if it can determine a rollback-point by choosing a proper activity as the start point of undoing, rather than undoing all the completed activities (aborting the running workflow instance). The following are the possible policies to resume the running workflow instance:

- *no rollback-points: kill-and-restart (abort) the running workflow instance with undoing all the completed activities*
- *rollback-points with randomly selected*
- *rollback-points with the most recently completed first*
- *predefined rollback-points: specified at the build-time (the modeling time) of a corresponding workflow process model*
- *the ideal policy: autonomous rollback-points with intelligently selected.*

Conclusively speaking, we try to dig out a kind of the ideal policies that ought to be appropriate for effectively as well as efficiently realizing our goal, keeping sustainable on workflow enactments, through a self-recoverable rollback mechanism. Therefore, in this paper, we newly propose a concept of rollback sequences that is based on the workflow risk dependency model formally described in [10]. The basic idea is this; we can reasonably assign a single rollback-point of gateway-activity, which is so-called "a risk dependency point," to every activity on a workflow process model, and the risk dependency point of each activity becomes a rollback point of the activity when it encounters with any type of exceptional situations. Note that the syntactical structure of a workflow process model is determined by the basic gateway primitives, such as disjunctive, conjunctive, and iterative gateway-activity types, and so three types of risk dependency points, such as alternative-risk, conjunctive-risk, and iterative-risk dependencies, can be deduced from a workflow process model, and they become the theoretical bases of a gateway-centered rollback-point ancestry model proposed in the paper.

3   Gateway-Centered Rollback-Point Sequences

We are proposing an autonomous rollback-points selection approach as one of the ideal policies, which is named as gateway-centered workflow rollback-points ancestry model. It is possible to deduce two types of rollback-points ancestry models by revising the algorithm [10] that is able to algorithmically generate a workflow risk dependency net [10] from an information control net [1,4], which are gateway-centered rollback-points ancestry model and task-centered rollback-points ancestry model. Our focus pays on the gateway-centered rollback-points ancestry model, in particular.

3.1   Information Control Net and Its Risk Dependency Model

In order to automatically generate a gateway-centered rollback-points ancestry model from a workflow process model, we would use the concept of workflow risk dependencies [10] that the authors' research group developed and presented in [10]. The concept of risk dependencies can be defined graphically and formally by generating the workflow risk dependency net and workflow risk dependency model, respectively, from an information control net of workflow process model. The concept of gateway-centered rollback-points ancestries can be deduced from a workflow risk dependency net of the corresponding information control net.

A workflow risk dependency net is for modeling the risky effects on control transition types, such as sequential, conditional (*or-split* and *or-join*), parallel (*and-split* and *and-join*), and loop (*loop-split* and *loop-join*) control transitions. In particular, the three types of gateway-centered control transitions (conditional, parallel, and loop) become important leverages to form three types of risk dependencies in a workflow risk dependency model by revising the risk dependency generation algorithm [10]. Assume that we have to use a series of special notations, operations and their meanings, which are closely related with the risk dependency analysis and precisely described in [10]. In terms of revising the algorithm, it is necessary to extend the domination-relationship operations [2,10] so as to incorporate the concept of gateway-centered risk dependency types, such as alternative-gateway, conjunctive-gateway, and iterative-gateway dependency types. However, we won't describe any further in this paper, because of the page limitation. In this paper we focus on formalizing the gateway-centered rollback-points ancestry model, in particular.

3.2   Gateway-Centered Rollback-Points Ancestry Model

As described in the previous, the novel concept of workflow rollback-points ancestries can be realized in either a gateway-centered rollback-points ancestry net or a task-centered rollback-points ancestry net, each of which can be automatically generated from a workflow risk dependency net [10] of an information control net of workflow process model. The former net is graphically representing a set of gateway-activities and their risk dependency edges, whereas the latter net is composed of a set of graphical nodes of task-activities and their risk dependency edges. In this section, we formalize the formal definition of gateway-centered rollback-points ancestry model, $M^g = (\chi^g, \vartheta^g, I, O)$, as shown in [Definition 1].

Definition 1. *Gateway-Centered Workflow Rollback-Points Ancestry Model from a workflow risk dependency model. Let M^g be a gateway-centered workflow rollback-points ancestry model, that is formally defined as $M^g = (\chi^g, \vartheta^g, I, O)$ over a set of gateway-activities, A^g, and a set of transition-conditions, T^g, where*

- $\chi^g = \chi^g_{re} \cup \chi^g_{rd}$
  where, $\chi_o : A^g \longrightarrow \wp(A^g)$ is a multi-valued mapping of a gateway-activity to an another set of gateway-activities, each member of which is a direct descendant having disjunctive, conjunctive, and iterative gateway-centered dependencies, and $\chi^g_{re} : A^g \longrightarrow \wp(A^g)$ is a single-valued mapping of a gateway-activity to an another gateway-activity that is a direct ancestor having disjunctive, conjunctive, and iterative gateway-centered effects.

- $\vartheta^g = \vartheta^g_{re} \cup \vartheta^g_{rd}$

  where, ϑ^g_{re}: a set of control-transition conditions, $\tau \in T^g$, on each arc, $(\varphi^g_{re}(\alpha), \alpha)$; and ϑ^g_{rd}: a set of control-transition conditions, $\tau \in T^g$, on each arc, $(\alpha, \varphi^g_{rd}(\alpha))$, where $\alpha \in A^g$;
- I is a finite set of initial input repositories;
- O is a finite set of final output repositories;

Figure 1 depicts the conceptual idea of the gateway-centered rollback-points ancestries, and how to use them in an autonomous rollback-point selection mechanism. The left-side of the figure is illustrating a simple information control net [3,9] having one pair of split-join alternative-gateway nodes and another pair of split-join conjunctive-gateway nodes, whereas the right-side illustrates a possible rollback-points sequence on the workflow risk dependency net of the corresponding information control net. In particular, $\chi^g = \chi^g_{re} \cup \chi^g_{rd}$ is formally representing the ancestral relationships with risky dependency and effect properties, where χ^g is among gateway-activities. These gateway-centered ancestral relationships are eventually used for choosing gateway-centered rollback-points of workflow instances spawned from a corresponding workflow process model.

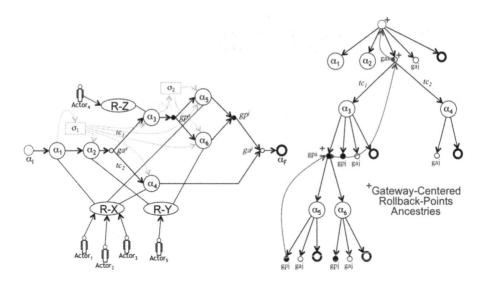

Fig. 1. An Information Control Net and Its Workflow Risk Dependency Model

In order to systematically construct a gateway-centered workflow rollback-points ancestry net from a workflow risk dependency net, the authors' research group has devised an algorithm that is named as the *"gateway-centered workflow rollback sequence ancestries generation algorithm."* The algorithm is able to produce a gateway-centered rollback-points ancestry net, such as "A Gateway-Centered Rollback-Points Ancestry Model ($M^g = (\chi^g, \vartheta^g, I, O)$)," from a workflow risk dependency net, "A Workflow Risk Dependency Net ($\Omega = (\varphi, \xi, I, O)$),"

The Gateway-Centered
Rollback-Points
Ancestry Net

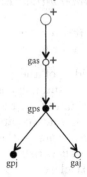

Fig. 2. The Gateway-Centered Workflow Rollback-Points Ancestry Net of Fig. 1

Table 1. Formal Representation of the Gateway-Centered Rollback-Points Ancestry Model for the Information Control Net of Fig. 1

$M^g = (\chi^g, \vartheta^g, I, O)$ over $\mathbf{A}, \mathbf{T}$ /* The Gateway-Centered Rollback-Points Ancestry Model

$\mathbf{A}^g = \{\alpha_I, g_a^s, g_p^s, g_p^j, g_a^j\}$ /* Gateway Activities

$\mathbf{T} = \{d(default), tc_1, tc_2\}$ /* Transition Conditions

$\mathbf{I} = \emptyset$ /* Initial Input Repositories

$\mathbf{O} = \emptyset$ /* Final Output Repositories

$\chi^g = \chi_{rd}^g \cup \chi_{re}^g$	$\chi_{rd}^g(\alpha_I) = \{\{g_a^s\}\};$	$\chi_{re}^g(\alpha_I) = \{\emptyset\};$
	$\chi_{rd}^g(g_a^s) = \{\{g_p^s\}\};$	$\chi_{re}^g(g_a^s) = \{\{\alpha_I\}\};$
	$\chi_{rd}^g(g_p^s) = \{\{g_p^j, g_a^j\}\};$	$\chi_{re}^g(g_p^s) = \{\{g_a^s\}\};$
	$\chi_{rd}^g(g_p^j) = \{\{\alpha_F\}\};$	$\chi_{re}^g(g_p^j) = \{\{g_p^s\}\};$
	$\chi_{rd}^g(g_a^j) = \{\{\alpha_F\}\};$	$\chi_{re}^g(g_a^j) = \{\{g_p^s\}\};$
$\vartheta^g = \vartheta_{rd}^g \cup \vartheta_{re}^g$	$\vartheta_{rd}^g(\alpha_I) = \{d\};$	$\vartheta_{re}^g(\alpha_I) = \{\emptyset\};$
	$\vartheta_{rd}^g(g_a^s) = \{\{tc_1, tc_2\}\};$	$\vartheta_{re}^g(g_a^s) = \{d\};$
	$\vartheta_{rd}^g(g_p^s) = \{d\};$	$\vartheta_{re}^g(g_p^s) = \{d\};$
	$\vartheta_{rd}^g(g_p^j) = \{\emptyset\};$	$\vartheta_{re}^g(g_p^j) = \{d\};$
	$\vartheta_{rd}^g(g_a^j) = \{\emptyset\};$	$\vartheta_{re}^g(g_a^j) = \{d\};$

---

described in [10]. Due to the page limitation, we won't describe the details of the algorithm. However, as an operational example, we have applied the algorithm to the workflow risk dependent net of Fig. 1, and its output, a gateway-centered rollback-points ancestry net, is graphically and formally arranged in Fig. 2 and Table 1, respectively.

In summary, we believe that the gateway-centered workflow rollback-points ancestry model plays a very important role as a risk recovery mechanism to realize the eventual goal of sustainable workflow process enactments. From the ancestral relationships' information of $\chi^g = \chi_{rd}^g \cup \chi_{re}^g$, we are able to build a series of autonomous rollback-point sequences for each of the workflow instances spawned from a workflow process model. In particular, χ^g_{re} and χ^b_{re} possess the hierarchy of gateway-centered ancestors that dominates every activities in a corresponding workflow process model. Therefore, we can automatically generate a series of rollback-point sequences of each workflow instance so as to be applied to recover from the risky situation where it is faced with. As an operational example, the following lists a series of rollback-point sequences that can be obtained from the gateway-centered ancestral relationships information, χ^g_{re} and χ^g_{rd}.

- Gateway-Centered Rollback-Points Sequence$_0$ = $g^j_a \rightarrow g^s_p \rightarrow g^s_a \rightarrow \alpha_I$
- Gateway-Centered Rollback-Points Sequence$_1$ = $g^j_p \rightarrow g^s_p \rightarrow g^s_a \rightarrow \alpha_I$

4   Conclusions

So far, this paper has deployed the concept of gateway-centered workflow rollback-points ancestries and its usage for autonomous workflow recovery mechanisms to achieve the sustainable workflow process enactment services. Based upon the theoretical concept, we are able to not only generate a series of gateway-centered workflow rollback-point sequences for a workflow process model, but also extract the gateway-centered rollback-points ancestral relationships to be used for the autonomous workflow recovery mechanisms. As stated previously, in resolving the sustainability issue of workflow process enactment services, it is very important for the system to provide autonomous error-detection and recovery functionalities on its running exceptions as well as very safe self-recovery mechanisms on the exceptional and risky situations. Therefore, we are certainly sure that gateway-centered rollback-point sequences from a gateway-centered rollback-points ancestry model proposed in the paper can produce valuable workflow risk dependency knowledge that will be eventually used for realizing the sustainable workflow processes enactment services.

References

1. Ellis, C. A.: Information control nets: a mathematical model of information flow. In: ACM Proceedings of the International Conference on Simulation, Modeling and Measurement of Computer Systems, pp. 225–240. ACM (1979)
2. Podgurski, A., Clarke, L.A.: A formal model of program dependencies and its implications for software testing, debugging, and maintenance. IEEE Trans. Software Eng. 16(9), 965–979 (1990)
3. Ellis, C. A., Nutt, G. J.: The modeling and analysis of coordination systems. University of Colorado/Dept. of Computer Science Technical report, CU-CS-639-93 (1993)
4. Ellis, C. A.: Formal and informal models of office activity. In: Proceedings of the 1983 Would Computer Congress, Paris (1983)
5. Kim, K., Ellis, C.A.: Workflow Reduction for Reachable-path Rediscovery in Workflow Mining. Series of Studies in Computational Intelligence, Foundations and Novel Approaches in Data Mining, Springer, Heidelberg (2006)

6. Kim, K.-H.: A process-driven inter-organizational choreography modeling system. In: Meersman, R., Tari, Z. (eds.) OTM-WS 2005. LNCS, vol. 3762, pp. 485–494. Springer, Heidelberg (2005)
7. Park, M., Kim, K.: Control-path oriented workflow intelligence analyses. J. Inf. Sci. Eng. **24**(2), 343–359 (2008)
8. Kim, K., Ra, I.: e-Lollapalooza: a process-driven e-business service integration system for e-logistics services. KSII Trans. Internet Inf. Syst. **1**(1), 33–51 (2007)
9. Kim, K.: A model-driven workflow fragmentation framework for collaborative workflow architectures and systems. J. Netw. Comput. Appl. **35**(1), 97–110 (2012)
10. Chun, M., et al.: Analyzing risk dependencies on rfid-driven global logistics processes. In: The Proceedings of the ACIS/JNU International Conference on Computers, Networks, Systems, and Industrial Engineering, pp. 59–64 (2011)
11. Maa, J., Wangc, K., Lida, X.: Modelling and analysis of workflow for lean supply chains. Enterp. Inf. Syst. **5**(4), 423–447 (2011)
12. Ghadge, A., Chester, M., Kalawsky, R.: A systems approach for modeling supply chain risks. Supply Chain Manag. Int. J. **18**(5), 523–538 (2013)

jBPM4S: A Multi-tenant Extension of jBPM to Support BPaaS

Dongjin Yu[1(✉)], Qi Zhu[1], Dalong Guo[1], Binbin Huang[1],
and Jianwen Su[2]

[1] School of Computer, Hangzhou Dianzi University, Hangzhou 310018, China
yudj@hdu.edu.cn
[2] Department of Computer Science, UC Santa Barbara, Santa Barbar, USA
su@cs.ucsb.edu

Abstract. BPaaS, or Business Process as a Service, is an advanced model of
SaaS in which the Business Process Management system is deployed as a hosted
service and accessed over the Internet without the need for the user to deploy
and maintain additional on-premise IT infrastructure. In this paper, we present
an architectural design and implementation of a BPaaS system, called jBPM4S.
jBPM4S is an extension of jBPM, and further provides process-related services
to be invoked by multiple tenants on their demands over the Internet. It leverages Spring Framework to manage transactions of process instance execution
and expose its universal process services that are unrelated to the specific
business through the means of Web services. The case study based on jBPM4S
demonstrates the effectiveness and efficiency of such service based process
management facility.

Keywords: Business process management · Business process as a service ·
jBPM · Multi-tenants

1 Introduction

Business process management (BPM) focuses on improving corporate performance
by managing and optimizing an enterprise's business processes. It enables flexible
and individual composition and execution of services as opposed to hard-coded
workflows in most off-the-shelf software [1]. In a traditional architecture, a BPM system
responsible for coordinating and monitoring running instances of business processes is
usually a part of enterprise application systems. However, to design, implement, and
maintain a BPM system is not always a straightforward task. Purchasing a BPM system
is an expensive investment for the enterprise. In addition, scalability and flexibility are
often a concern for enterprises that use such "in-house" BPM systems, since a process
engine is only able to coordinate a limited number of business process instances
simultaneously [2].

Cloud computing has changed how computing, storage, and software services are
provisioned. It gives its users the opportunity of using computing resources as common
utility in a pay-per-use manner, even with a perception that these resources are
unlimited, always and instantly available [3, 4]. As we all know, cloud computing

J. Bae et al. (Eds.): AP-BPM 2015, LNBIP 219, pp. 43–56, 2015.
DOI: 10.1007/978-3-319-19509-4_4

providers offer three major system services: Software as a Service (SaaS), Platform as a Service (PaaS), and Infrastructure as a Service (IaaS). In SaaS, consumers pay for a software subscription and move all or part of their data and the managing code to remote servers [5]. Business Process as a Service, or BPaaS, is a new paradigm of SaaS, which can be described as deployed as a hosted business process management service and accessed over the Internet without the need for the user to deploy and maintain additional on-premise IT infrastructure [6]. From a BPaaS vendor's perspective, the benefits of BPaaS arise from leveraging economies of scale and separating the process management from domain business, by serving a large number of customers ("multiple tenants") through a shared, centrally-hosted software service for managing business processes.

In this paper, we present the design and implementation of a prototype of BPaaS based on jBPM, a world-leading open source project for business process management. The prototype system, called jBPM4S, provides the services of business process management over the Internet for different applications and users. Through the aid of jBPM4S, traditional enterprise applications no longer need to embed a workflow engine to fulfill the process-related tasks. In addition, when jBPM4S is deployed and operating over the Internet, multiple enterprises/organizations can rent the process management-related services instead of installing and maintaining their own BPM systems. As a dedicated BPM system for multi-tenants, jBPM4S can be deployed on the cloud-based IT infrastructure services, such as Amazon Web Services (AWS) and Aliyun, the latter lacking the support for understanding processes and process management functions if they are used alone.

This paper is structured as follows. We discuss some background information to motivate our work in Sect. 2. We then introduce the design of BPaaS system on the Cloud and how the tenants, or application systems, interact with the BPaaS system in Sect. 3. In Sect. 4, we describe the detail implementation of jBPM4S that is a prototype of BPaaS system. A case study and its running performance are illustrated in Sects. 5 and 6 to show the effectiveness of jBPM4S. Following the discussions on related work in Sect. 7, we summarize our work and outline a few future research directions in the final section.

2 Motivations

A business process consists of an assembly of activities, which are performed by either humans or information (software) systems. The process links together people, information flows, software and other systems, and other assets to create and deliver value to customers. In a Business Process Management System (BPMS) that has been in existence for decades, the process execution engine is responsible for coordinating and monitoring running instances of business processes. Consider as an example an *absence approval* process in a (Chinese) college. Usually, a student who wants a temporary leave (e.g., for a family emergency) from school when school is in session should submit her absence request beforehand and get it approved. Only after her request is approved, she can then leave the campus. In addition, if the leave should last longer than the approved period, she would need an extension approved by the dean.

This absence approval service illustrates a simple business process. To facilitate the approving process for potentially numerous absence requests in a large college, the absence approval process should be designed and demand help from a software system that is to be developed. An embedded or standalone BPMS could used to coordinate and administrate the instances of reviewing absence requests.

If more than one college would deploy the absence approval workflow system, to migrate the *Absence Approval System* to the Cloud as a SaaS application would be a desirable approach, as Fig. 1 depicts. In this way, the colleges pay for a subscription to use *Absence Approval System* instead of design, implement, and maintain their own ones. The colleges, as the tenants, should benefit from the advantages of cloud computing: cost reductions, pay-as-you-go pricing models, quick time to market and economies of scale.

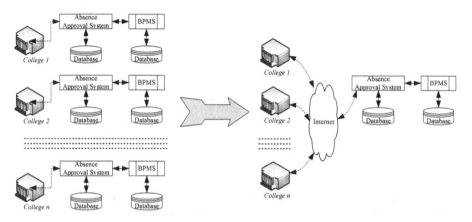

Fig. 1. To migrate the application systems to the cloud as a SaaS application

However, while this appears nice, there are difficult problems when many colleges share the same *Absence Approval System*. In the setting, different colleges are served simultaneously by the same installation of the *Absence Approval System* (in the cloud) while features of the system should remain logically separated for different colleges. Specifically, when the reviewing rules differ slightly or substantially, customization for each college is required as an architectural feature provided by the *Absence Approval System* in the cloud to facilitate efficient and faithful incorporation and management of college-specific requirements. An effective and efficient BPaaS model is illustrated in Fig. 2(a), in which the BPMS is deployed in the cloud whereas the *Absence Approval Systems* are still local to each college. In other words, BPMS provides process execution and management as services for colleges (clients). This BPaaS approach has the following four advantages at least. First, it reduces IT support costs at the clients' sides by outsourcing hardware and software maintenance and support to the BPaaS provider. This advantage comes naturally from the delivery model of SaaS. Second, it separates the business process engine from the concrete and specific application system(s), which makes the maintenance of application systems simpler, and thus more convenient and productive. This advantage is a consequence of using BPMS. Third, a BPMS is possible to be harnessed to its maximum capacity because it can be tailored to serve multiple application systems. As Fig. 2(b) indicates, the BPMS in the cloud provides process services for both *Absence Approval System*s and *Reimbursement Systems* for a couple of colleges. Last but not least, the fears concerning security and privacy issues are minimized to the least extent since the application systems, including its business data and business logic, are still controlled by the clients who are actually tenants of BPMS.

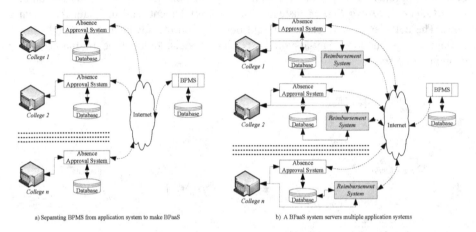

a) Separating BPMS from application system to make BPaaS   b) A BPaaS system servers multiple application systems

Fig. 2. To migrate BPMS to the Cloud as a BPaaS application

There are a dozen of open-source BPM systems available around the world, such as jBPM (www.jbpm.org), YAWL (www.yawlfoundation.org), and Activiti (www. activiti.org). Unfortunately, none of these open-source products supports the BPaaS model described in the above discussions. Moreover, the traditional approach to adoption of jBPM and Activiti in the development is to embed the BPMS in the application system(s). As for YAWL, its service-oriented architecture merely means one can replace or extend existing components with newly developed ones.

In summary, the problem this paper intends to address is as the following: *How to design and implement a business process management system that can be deployed with the BPaaS model*. Since developing such system from scratch is a huge and long-term job and jBPM is regarded as a popular and world leading open-source BPM system in software development, we tackle the problem with the following approach: *Extend jBPM to provide the business process service over the Internet*.

3   Architectural Design of BPaaS System

This section outlines the design of the BPaaS system: Sect. 3.1 presents its architecture and explains the key components, and Sect. 3.2 illustrates how it interacts with multiple business processes from multiple tenants.

3.1   The BPaaS System on the Cloud

Figure 3 shows the logical architecture for the BPaaS system and how it supports services of business process management for multiple tenants. The components of BPaaS system, illustrated in red rectangles in Fig. 3, are spread in the cloud and on the tenant side. The BPaaS system in the cloud (BPaaS server) employs the process management engine for requests of all user-defined business process services, such as process execution, process administration and process audit. On the other hand, the application systems, or the tenants, interact with the BPaaS system that provides process-related services over the Internet. The components of the BPaaS system are described in detail in the following.

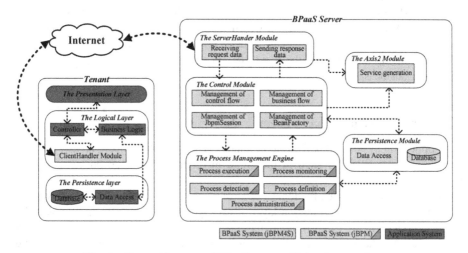

Fig. 3. The architecture of BPaaS system (Color figure online)

The ServerHandler Module. The *ServerHandler* module is mainly used to receive the service request from the external enterprise application systems, or tenants in other words, and return the response. Here, the request includes the tenant ID, the process instance ID and the business data, whereas the response provides the results of the process execution, such as the execution status and other running information.

The Axis2 Module. The *Axis2* module deals with the service generation based on Apache Axis2 (axis.apache.org). It encapsulates and publishes the capabilities pertain to the management of business process to the public. More specifically, it provides the services interfaces for all users, such as the service names, the required inputs, and the expected returning data type, etc.

The Process Management Engine. The *Process Management Engine* is the core module of BPaaS system. It models the processes in the BPMN language, executes process instances, manages the rules of processes, and monitors the status of process execution.

The Persistence Module. Based on user persistence strategy, the *Persistence* module stores process execution status, process execution results and the other related data to the relational database system. Since the process data are identified with the tenant IDs, the tenant IDs are used to easily separate data of one tenant from that of others.

The Control Module. The *Control* module ensures the correct information circulation among the different components. It guides the tasks before and after sending and receiving the process data that the *ServerHandler* module deals with, manages the transactions of the business process engine, and controls the operation of the persistent data. In addition, it controls the access from different tenants based on their predefined access rights.

The ClientHandler Module. The *ClientHandler* is the only module located on the tenant side. It is the entry point from which the tenant requests the BPaaS server for process-related services. In a typical Model-View-Controller application system, the *ClientHandler* is managed by the *Controller* in the logical layer for the interactions with local business logics.

3.2   Tenants Calling BPaaS System

The application systems play the role of tenants. They call the process-related services provided by the BPaaS system in the cloud in order to fulfill the business process logics.

More specifically, when the logical control layer of application system receives the request issued from its presentation layer, it packages the request data and invokes the process service via the *ClientHandler* module. After the *ServerHandler* on the BPaaS server receives the request, it decodes the request data and forwards them to the *Control* Module. The *Control* Module then delegates the request to the *Process Management Engine* which actually executes the process-related task as required. Only after the BPaaS system has dealt with the request and returned the results via the *ServerHandler*, the operation is considered to have been completed. A real example of executing process instances is given in Sect. 5.

According to [6], the execution of business process instances needs at least the following five types of data: (i) business data for the process logic, (ii) business process models, (iii) execution states (and histories), (iv) correlations among business process instances, and (v) resources and their states. In our solution, the data of type (i) and (v) are persisted in the application system on the tenant sides, whereas the data of type (ii) and (iv) are kept in the BPaaS system in the cloud via the *Persistence Module*. Meanwhile, the data of type (iii) are duplicated from the BPaaS system to the application system.

To create and execute business processes of tenants based on the BPaaS system mainly includes the following two steps: (1) Process modeling, which involves defining the start and end nodes of the business process, establishing the activity nodes, determining the sequences and the relationship among activity nodes and specifying the participating roles for activity nodes. (2) Form definition, which involves designing the user interfaces for human tasks and calling the process-related Web services provided by the BPaaS system. Here, a form can be roughly mapped to its corresponding artifact. In other words, the operation and administration of the process instances are fulfilled on the Cloud by the BPaaS system, although the concrete business logics are mainly realized by the application system itself.

4 jBPM4S: The Implementation of BPaaS System

We developed a prototype of BPaaS system, called jBPM4S, based on the jBPM. We extend jBPM (the components in red rectangles with blue triangles attached in Fig. 3) to a BPaaS system (the components in red rectangles in Fig. 3) so as to provide the process-related services for the multiple tenants which deal with business processes. The advantages of jBPM4S can be highlighted in the following two aspects. (1) It leverages Spring Framework to manage the transaction of process instance execution so that the business and process operations would be executed in the same transactions. (2) It exposes its universal process services that are unrelated to a specific business through the means of Web services. In this way, jBPM4S can be a "bare bone" and non-intrusive service provider to the concrete applications.

4.1 Integration of jBPM and Spring Framework

First of all, we need to integrate jBPM with the transaction control component of Spring Framework (spring.io), one of the widely-used traditional Model-View-Controller frameworks. When jBPM is in operation, the business operation and the process operation must be executed in the same transaction. Take the process of absence approval as the example. When the absence request is submitted, the executing process instance jumps to the next task node and saves the information of application in the database of tenant side. Here, these two operations need to occur in the same transaction. Otherwise, it may cause the conflicts or inconsistencies between business and process data.

Therefore, when developing a jBPM4S, the business data need to be manipulated while the process operation is performed. However, jBPM controls the operation of processes by its core engine. On the other hand, Spring Framework controls the business logic operations by itself. Figure 4 shows how we integrate jBPM with Spring Framework, where Fig. 4(a) suggests jBPM and Spring Framework execute in the different transactions, and Fig. 4(b) suggests both execute in the same transaction. Thus, the processing flow could be illustrated as follows. Firstly, the *BeanFactory* provided by the Spring Framework initializes the configuration of jBPM and then creates a *StatefulKnowledgeSession* object. Afterward, with the help of *sessionFactory*, the business logics are implemented and the processes are executed. Finally, both processes operations and business logic operations call *Hibernate* for persistence operations.

We use the configuration file named *application-context.xm*l to integrate jBPM and Spring Framework. The core of the configuration is as follows.

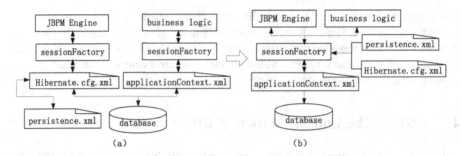

Fig. 4. The integration of jBPM and Spring Framework to manage the transactions of process execution.

```
<bean id="jbpmEntityManagerFactory"
  class="org.springframework.orm.jpa.LocalContainerEntity
  ManagerFactoryBean">
  <property name="dataSource" ref="xDataSource" />
  <property name="persistenceUnitName"
    value="org.jbpm.persistence.local" />
</bean>
<bean id="jbpmTxManager" class
  ="org.springframework.orm.jpa.JpaTransactionManager">
  <property name="entityManagerFactory"
  ref="jbpmEntityManagerFactory" />
  <property name="nestedTransactionAllowed"
    value="true" />
</bean>
<drools:ksession id="ksession" type="stateful"
  kbase="kbase" node="node1">
  <drools:configuration>
    <drools:jpa-persistence>
      <drools:transaction-manager ref="jbpmTxManager" />
      <drools:entity-manager-factory
        ref="jbpmEntityManagerFactory" />
    </drools:jpa-persistence>
  </drools:configuration>
</drools:ksession>
<bean id="myLocalTaskService" class
  ="com.gds.jbpm.MyLocalTaskService">
  <property name="ksession"    ref="ksession" />
  <property name="localTaskService"
    value="taskService" />
</bean>
```

4.2   Interfaces of Process-Related Services

The second challenge comes from the design and implementation of process-related service interfaces. Since jBPM4S provides process-related services for different application systems (tenants) no matter what these systems deals with, the interfaces should be designed to accommodate to the general usage. We design two standard interfaces, presented as follows, in which both the input and output are irrelevant to the concrete applications.

```
//to initiate the process execution
public DataReturned initiateProcessInstance(String tenantID
  , String processID, DataBean dBean)
    Input:
      tenantID: the tenant ID
      processID: the process ID to be initiated
      dBean: business data packet that would be resolved by
        jBPM4S for process execution, such as number of days
        off.
    Output:
      business data packet responding from jBPM4S, including the
        current process instance ID, the following task node
        ID, the following task performer, execution status,
        etc.

//to continue the execution of process instance
public DataReturned go-forward(String tenantID, String
  processInstanceID, DataBean dBean)
    Input:
      tenantID: the tenant ID
      processInstanceID: the process instance ID to be executed
      dBean: business data packet that would be resolved by
        jBPM4S for process execution.
    Output:
      business data packet responding from jBPM4S, including the
        following task node ID, the following task performer,
        execution status, etc.
```

Here, *dBean*, as one of the input parameters, is a map containing a list of *key-valued* business data. jBPM4S parses *dBean* and matches the business data one by one with the required input of task node based on their keys. The applications are then able to complete the execution of business processes or transfer the process instances to humans as human tasks. In this way, the applications, no matter what they are, can customize their own processes and send their specific data to jBPM4S on the Cloud by calling the service interfaces. Meanwhile, jBPM4S can automatically choose the related process instances to execute according to tenant ID, process ID and process instance ID.

5   A Case Study

We implemented jBPM4S, a prototype of BPaaS system, based on the design given above. Here, we use the process of absence approval to show how it works. The process is defined by BPMN, whose graphic notation is illustrated in Fig. 5.

Suppose one student who wants to apply the day offs logs in *Absence Approval System*. She fills in the application form. When she submits the form, she initiates a new process instance for absence approval. In other words, a request of *initiatePro-cessInstance* together with the tenant ID, the process ID, and the number of absence days, is sent to jBPM4S on the Cloud. jBPM4S then parses the request and takes the appropriate action. When finishing, it returns the results back to *Absence Approval System*.

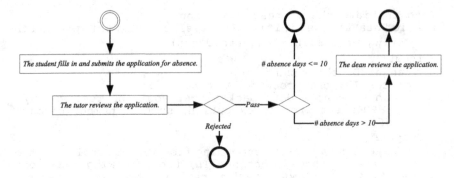

Fig. 5. The process definition of absence approval application

Afterward, the tutor logs in *Absence Approval System* and finds the absence request as Fig. 6 presents. If the tutor approves the absence request, *Absence Approval System* calls the *go-forward* interface. Then, the business data are packaged and sent back to jBPM4S. In other words, *Absence Approval System* does not need to care about the rules of execution, and just waits for the execution results of related the process instance.

Absence Approval System

Current User:gdl

Welcome gdl !	Applicant	Reasons	Days	State	Date	Process ID
	gdl	A medical ...	7	Pass	2015-01-25	9
Application from Student	gdl	Application for...	10	Pass	2015-01-25	10
	gdl	I got a bad cold...	2	Reject	2015-01-25	11
Approval by Teacher	gdl	I want to ask for ...	15	Waiting for the dean review	2015-01-25	12

Approval by Dean

Login

Fig. 6. The tutor approves the absence request

Finally, the dean logs in *Absence Approval System* and finds the absence request as Fig. 7 presents. Here, the duration of absence requests are all equal to or longer than ten days. If the dean approves one request, *Absence Approval System* calls the *go-forward* interface. Similarly, the business data are packaged and sent back to jBPM4S.

In order to verify the effectiveness of jBPM4S for serving multiple application systems, we also migrate the execution and administration of reimbursement processing from *Financial Reimbursement System* to jBPM4S. Similarly, *Financial Reimbursement System* initiates the reimbursement process instance through *initiate-ProcessInstance* interface, and continues the execution of instance through *go-forward* interface. Although these two application systems are totally different, they do not interfere with each other simply because the business data are managed and mainly persisted on the tenant sides. Under such circumstance, jBPM4S just focuses on the status of running process instances.

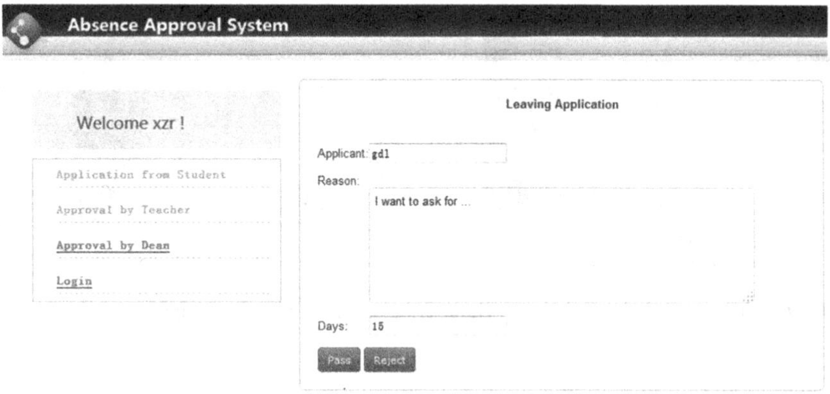

Absence Approval System

Welcome xzr !

Application from Student

Approval by Teacher

Approval by Dean

Login

Leaving Application

Applicant: gd1

Reason:

I want to ask for ...

Days: 15

Pass   Reject

Fig. 7. The dean approves the absence request

6 Performance Evaluation

In order to evaluate the performance of jBPM4S, we deployed jBPM4S on a Windows 7 64-bit server with Intel Core i3-2120 CPU and 4 GB memory. During the experiment, we employed Apache JMeter (jmeter.apache.org), a software designed for load testing, to initiate the requests to jBPM4S and measure the response time and throughput. Figure 8(a) demonstrates the average and maximum response time in milliseconds to the number of requests initiated concurrently in 10 s, whereas Fig. 8(b) demonstrates the throughputs per second. As indicated, both the response time and the throughput were increased if more requests were initiated. However, the throughput was dramatically decreased when jBPM4S was overloaded at receiving 20 requests in 10 s.

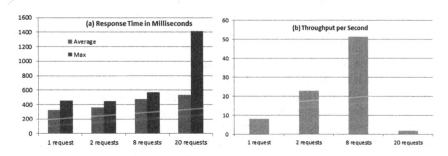

Fig. 8. The response time and throughput to the number of requests

To investigate how different types of applications impact on the performance of jBPM4S, we measured the response time and throughput for tenants of just one type of application and that of two types of applications. As indicated in Fig. 9, the performance for tenants of mixed applications was much better than that of single application. The reason is probably that the tenants of different applications had different behaviors and staggered their requesting periods. It is just what BPaaS is expected to bring about.

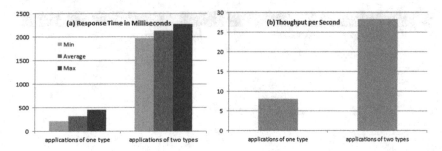

Fig. 9. The response time and throughput for one and two types of application systems

7 Related Work

In recent years, BPaaS has attracted the attention of both academia and industry. There have been a number of approaches for BPaaS presented in the literature. In [2], Duipmans et al. introduce a transformation-based approach that allows companies to control the parts of their business processes that should be allocated to their own premises and to the Cloud, to avoid unwanted exposure of confidential data and to profit from the high performance of cloud environments. Although outsourcing the execution of configurable business processes could gain a vast economic potential, both external and internal management pose a general challenge for its acceptance. In [7], Accorsi reviews the role of remote auditing as a means to address this issue and indicates research directions for automated tool support. Besides, Muthusamy et al. present a Service Level Agreements-driven approach to business process management for service-oriented applications in environments such as Cloud computing platforms, as described in [8]. On the other hand, multi-tenant applications are often limited in terms of customizability since one application should fit the needs of all customers. In [9], Gey et al. focus on the customizability due to the different tenant-specific requirements. They argue that the workflow modelling language should offer support to describe these tenant-level variabilities explicitly. Moreover, Dana Petcu et al. propose an approach to the applicability of re-usable business process patterns and a multi-modeling approach for BPaaS design, deployment and operation [10].

Although BPaaS has been widely accepted, its implementation is seldom reported. In [6], Sun et al. formulate a concept of a self-guided artifact, which extends artifact-centric business process models by capturing all needed data for a business process throughout its execution. Similarly, Zheng et al. describe a business process oriented Platform-as-a-Service framework called BPPaaS, which includes an integrated business process application programming model, and a business process oriented Platform-as-a-Service middleware [11]. jBPM4S presented in this paper, however, is somewhat different from that reported in the literature, in which it leverages Spring Framework to manage the transaction of process instance execution. In addition, jBPM4S exposes its universal process services which are unrelated to a specific business through the means of Web Services to be invoked by multiple tenants over the Internet.

8   Conclusions

Application-level multi-tenancy is an increasingly prominent architectural pattern in Software-as-a-Service (SaaS) applications that enables multiple tenants (customers) to share common application functionality and resources among each other. As one of the instances of SaaS model, the provision of Business Process as a Service (BPaaS) grows quickly, especially for small and medium enterprises that cannot afford fully-fledged enterprise solutions. In this paper, we provide a prototype of BPaaS system, called jBPM4S, which is extended from jBPM. jBPM4S provides the process-related services over the Internet for multiple tenants. The application systems, as the tenants, call jBPM4S for process execution and administration on their demands.

The current version of jBPM4S holds the process information about multiple tenants, such as execution status and process models. In the future, we will investigate how to isolate process information about one tenant from these of others to further improve the safety of multi-tenancy. Moreover, we plan to deploy jBPM4S on the AWS or Aliyun to test its performance while dealing with a large number of users in real applications.

Acknowledgments. The work is supported by National Natural Science Foundation of China (No. 61100043), Zhejiang Provincial Natural Science Foundation (No. LY12F02003), the Key Science and Technology Project of Zhejiang (No. 2012C11026-3).

References

1. Stefan, S., Janiesch, C., Venugopal, S., Weber, I., Hoenisch, P.: Elastic Business Process Management: State of the art and open challenges for BPM in the cloud. Future Generation Computer Systems. http://dx.doi.org/10.1016/j.future.2014.09.005 (2014, in press)
2. Duipmans, E.F., Pires, L.F., da Silva Santos, L.O.B.: A transformation-based approach to business process management in the cloud. J Grid Comput. **12**(2), 191–219 (2014)
3. Bibi, S., Katsaros, D., Bozanis, P.: Business application acquisition: on-premise or SaaS-based solutions? IEEE Softw. **29**(3), 86–93 (2012)
4. Cusumano, M.: Cloud computing and SaaS as new computing platforms. Commun. ACM **53**(4), 27–29 (2010)
5. Sengupta, B., Roychoudhury, A.: Engineering multi-tenant software-as-a-service systems, In: Proceedings of the 3rd International Workshop on Principles of Engineering Service-Oriented Systems, Co-located with ICSE 2011, pp. 15–21 (2011)
6. Sun, Y., Su, J., Yang, J.: Separating execution and data management: a key to business-process-as-a-service (BPaaS). In: Sadiq, S., Soffer, P., Völzer, H. (eds.) BPM 2014. LNCS, vol. 8659, pp. 374–382. Springer, Heidelberg (2014)
7. Accorsi, R.: Business process as a service: chances for remote auditing. In: 35th IEEE Annual Computer Software and Applications Conference Workshops, pp. 398–403 (2011)
8. Muthusamy, V., Jacobsen, H.-A.: BPM in cloud architectures: business process management with SLAs and events. In: Hull, R., Mendling, J., Tai, S. (eds.) BPM 2010. LNCS, vol. 6336, pp. 5–10. Springer, Heidelberg (2014)

9. Gey, F., Walraven, S., Van Landuyt, D., Joosen, W.: Building a customizable business-process-as-a-service application with current state-of-practice. In: Binder, W., Bodden, E., Löwe, W. (eds.) SC 2013. LNCS, vol. 8088, pp. 113–127. Springer, Heidelberg (2013)

10. Petcu, D., Stankovski, V.: Towards cloud-enabled business process management based on patterns, rules and multiple models. In: 2012 10th IEEE International Symposium on Parallel and Distributed Processing with Applications (ISPA), pp. 454–459 (2012)

11. Zheng, Y.Q., Pang, J.S.: Business process oriented platform-as-a-service framework for process instances intensive applications. In: 26th IEEE International Parallel and Distributed Processing Symposium (IPDPS), pp. 2320–2327 (2012)

Resources Allocation Strategies

An Efficient Resource Allocation Mechanism Based on Dynamic Pricing Reverse Auction for Cloud Workflow Systems

Xuejun Li[1,2], Xiangjun Liu[1], and Erzhou Zhu[1(✉)]

[1] School of Computer Science and Technology, Anhui University, Hefei, China
xjli@ahu.edu.cn
[2] Key Laboratory of ICSP, Ministry of Education,
Anhui University, Hefei, China

Abstract. The reverse auction method is widely applied to resource allocation for cloud workflow systems, and it dynamically allocates resources depending on the supply and demand of market. However, during the auction the price of cloud resource is usually fixed, and the resource allocation mechanism cannot adapt to the changeful market. This results in low efficiency of resource utility. In this paper, we first propose a dynamic pricing strategy in reverse auction, and then present an efficient resource allocation mechanism based on dynamic pricing reverse auction. During the auction, providers change resource price according to the trade situation so that the novel mechanism can increase chances of making a deal and improve efficiency of resource utility. In addition, providers improve their competitiveness by lowering price. Simultaneously users are timelier to get resources with the increasing of trade rate. Therefore, users can obtain cheaper resources in shorter time, which decrease monetary cost and completion time for workflow execution. The experiments show that our mechanism can achieve better results in resource utility, monetary cost and completion time compared to BOSS algorithm.

Keywords: Cloud computing · Reverse auction · Pricing · Resource allocation · Workflow

1 Introduction

Cloud Computing is a specialized distributed computing paradigm, it differs from traditional ones in that (1) it is massively scalable, (2) it can be encapsulated as an abstract entity that delivers different levels of services to customers outside the Cloud, (3) it is driven by economies of scale, and (4) the services can be dynamically configured and delivered on demand [1].

Workflow consists of a set of tasks and these tasks have partial order relation [2]. It is often required to be executed in a distributed high-end computing environment, so it is common to deploy workflow in the cloud environment. Cloud computing resources assigned to a workflow can be dynamically changed at runtime and the type and number of computation resources are determined by service requests [3] and resources can be assigned only when they are needed.

© Springer International Publishing Switzerland 2015
J. Bae et al. (Eds.): AP-BPM 2015, LNBIP 219, pp. 59–69, 2015.
DOI: 10.1007/978-3-319-19509-4_5

Cloud computing environment can provide many types of resources and the requirements of cloud workflows are also various, so resource allocation of cloud workflows is very important. Allocating cloud resources to users is a non-trivial task which is an NP-hard problem [4]. Resource allocation mechanism is generally based on conventional models and economic and game-theoretic models [5, 6]. Conventional model requires global knowledge and complete information. These mechanisms are mostly centralized in nature. The cost model of the centralized mechanism derives cost based on the usage of the resources. Economic model of resource management can offer incentives to participants. These models decide cost based on the value that the user gets from the services.

Different from conventional method, market-oriented method assumes that resource providers and users are rational and intelligent [6]. And the allocation depends on many factors including supply and demand and price of resources. Auction is a powerful tool to allocate resources in market. Formally speaking, auction is a protocol that allows participants to indicate their interests in different resources and use these indications of interest to determine both resource allocation and cost [5]. Reverse auction method is a type of auction. In general auction, there are one seller and multiple buyers. While in reverse auction, there are multiple sellers and one buyer in opposite way [7]. In reverse auction, the user sends the specification to the cloud broker and requests resources. The cloud broker transfers the user specification to all cloud providers. The cloud provider sells proper resources with price and ability. And then the user selects the optimal resources according to the criteria [8, 9]. In the most relevant literature [9], the author proposes BOSS mechanism based on reverse auction to allocate resources for tasks of workflows and each task starts an auction and gets a resource to minimize the monetary cost and completion time. However the price of resource is fixed during the auction, so that some providers with weaker competitiveness may lose the auction all the time and their resources never make a deal. This leads to low efficiency of resource utility.

In this paper, we firstly present a dynamic pricing strategy to change resource price according to trade situation, then propose a dynamic pricing-based allocation mechanism in order to improve efficiency of resource utility. It is common that computation ability of resource is difficult to improve in a short time. So changing the price of resource is a usual and efficient way to increase competitiveness. In this paper, this approach is adopted. In the dynamic pricing strategy, the resource price changes according to trade situation. These providers who win an auction keep resource price unchanged because they already have high competitiveness. But for those who lose an auction, decreasing the resource price in certain rate is an effective way to increase chances of making a deal. This strategy improves efficiency of resource utility, because the resources of providers with weaker competitiveness are sold. In addition, because resource utility increases, workflows execute timely and completion time decreases. In dynamic pricing strategy providers change their resource price to make resource be more cost-effective to attract more users. Therefore, the mechanism can decrease monetary cost.

Finally, the experiment results show that our mechanism can achieve a better result in resource utility, monetary cost and completion time compared to BOSS. Our main contributions are summarized as follows:

(i) Dynamic pricing strategy in reverse auction is proposed to change the resource price according to trade situation during the auction. Compared with traditional fixed pricing strategy, our strategy changes price to improve competitiveness according to providers' trade situation. So that providers win auction more easily and increase their revenue.

(ii) A dynamic pricing-based allocation mechanism is presented to allocate resource for workflow systems. Providers change resource price to sell more resources according to dynamic pricing strategy especially for those whose competitiveness is weak. This leads to higher resource utility, because more resources are sold. For users, more cost-effective resources appear in market because of competition of providers. So they can obtain the cheaper resource timely and therefore decrease completion time and monetary cost.

The rest of the paper is organized as follows. The Sect. 2 describes some related research. In Sect. 3, we show an example to analyze the problem. Section 4 proposes a dynamic pricing strategy, and Sect. 5 presents dynamic pricing-based allocation mechanism. In Sect. 6, traditional mechanism and ours are evaluated. Finally, Sect. 7 concludes the paper and discusses our future work.

2 Related Work

Resource allocation has received much attention from the research community as it is a crucial problem of large-scale distributed systems. Wood et al. [10] proposed an approach for dynamic allocation of resources by defining a unique metric based on the consumption of the three resources: CPU, network and memory. Gorlach and Leymann propose a method for dynamic provisioning of services in clouds in order to optimize the distribution of services [11]. However their proposed approaches can't satisfy requirements of users flexibly because the allocation of resource is pre-scheduled.

Market-based resource allocation has received much attention as it is a significant problem of large-scale distributed systems. In [12], authors present a model for resource allocation in grid using economic-based concepts including commodity market, posted price modeling, contract net models bargaining modeling, etc. Ludwig et al. present a heuristic program for resources allocation on utility computing infrastructure [13]. This heuristic program optimizes the number of resources allocated to tasks of workflow and speed up the execution within a limitation of budget. However these methods don't focus on the pricing mechanism of the resource allocation.

Pricing of resources is an important aspect in the field of resource allocation. In [2], authors mention Commodity Market which means that the sellers set the price for merchandise and the buyers pay money to get it. The price is pre-determined by the seller and does not change over time based on the balance of supply and demand. But the fixed pricing is not suitable for the changing market of cloud resource.

Dynamic pricing in the cloud has gained wide attention from both industry and academia. Amazon EC2 has introduced a "spot pricing" scheme for its resource instances where the spot price is the equilibrium price that determined according to resource demand and supply. To capture the realistic value of the cloud resources, the authors employ a financial option theory and treat the cloud resources as real assets [14].

The cloud resources are then priced by solving the finance model. In [15], the authors study the case of a single provider operating an IaaS cloud with a fixed capacity, while authors in [16] focus on the case of an oligopoly market with multiple providers. However, both [15, 16] give premise that the user's resource request is relying on resource price, which means that users reduce the demands of resources when price increase. This premise is not rational when users have executing deadline or cost constraint because users have to request the required amount of resources no matter what the price are.

In [17, 18], the authors present auction-based mechanisms to determine optimal resource price, taking into account the user's budgetary and deadline constraints. However, they consider the pricing model of only one provider. In [8] the authors assume that cloud vendors are selfish (interested in maximizing their own profits) and rational (able to appropriately calculate values and derive choices based on available information). And then present a cloud resource allocation approach based reverse auction which not only selects suitable cloud resource providers for users but also implements dynamic pricing. In [9], authors introduce a pricing model and a truthful mechanism for scheduling single tasks considering two objectives: monetary cost and completion time which are based on reverse auction. However, they don't consider about the competitiveness among providers, this leads to a consequence that the losers may always be losers because they don't try to improve their competitiveness. To solve this problem, we propose dynamic pricing-based scheduling mechanism for allocating resources efficiently, in which providers who lose this auction will decrease resource price for gaining more profit.

3   Problem Analysis

In this section, an illustrative example is given to explain the difference of pricing mechanism between BOSS algorithm and ours.

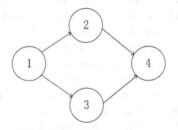

Fig. 1. Workflow example

Table 1. Time and cost of two resources

Task	R1		R2	
	Time	Cost	Time	Cost
1	4	3	2	5
2	6	6	4	7
3	6	6	4	7
4	2	2	1	3

In Fig. 1, there is an example workflow containing four tasks. The execution order of tasks depends on partial order relation of tasks and the child tasks start only when their parent tasks finished. For simplicity, we make the order of scheduling be $\{1, 2, 3, 4\}$. Table 1 indicates the initial time and cost of resources for executing each task. For BOSS algorithm [9], tasks start an auction according to the specific order to select the resource with the minimum product of time and cost. And provider gives its bid $bid_{i,j} = (t_{i,j}, c_{i,j})$ which means that the resource j completes the task i with time $t_{i,j}$ and cost $c_{i,j}$. The workflow makespan is the time required for executing the whole workflow under the partial order relation of tasks and the total cost of workflow is cost sum of all tasks. However, the price of each resource is fixed all the time even though the resource always loses. By calculation, the workflow makespan is 10 and the total cost of workflow is 20 (the calculation detail is presented in [9]).

In our mechanism, the difference is that the resource price is dynamic considering the competition among providers. If a provider wins this auction, he keeps resource price unchanged. However if a provider loses this auction, he decreases his resource price at a certain ratio in next auction. Here the price reduction ratio is 10 %.

In the first auction, the start time is assumed to be 0 and resources R_1 and R_2 give the bids $bid_{1,1} = (4, 3)$, $bid_{1,2} = (2, 5)$ respectively. Task 1 chooses $bid_{1,2}$, because $(2 + 0) \times 5 < (4 + 0) \times 3$. So the winner is R_2., and the completion time is 2, the cost is 5. For the loser R_1, it will decrease resource price by 10 % and the results are $bid_{2,1} = (6, 5.4)$, $bid_{3,1} = (6, 5.4)$, $bid_{4,1} = (2, 1.8)$.

Similarly, in the second auction, the winner is R_2, and the completion time is $2 + 4 = 6$, the cost is 7. For the loser R_1, it will continue decreasing the resource price and the results are $bid_{3,1} = (6, 4.86)$, $bid_{4,1} = (2, 1.62)$. In the third auction, the winner is R_1., the completion time is $2 + 6 = 8$, and the cost is 4.86. For the loser R_2., it decreases the resource price and result is $bid_{4,2} = (1, 2.7)$. In the last auction, the winner is R_1, and the completion time is $8 + 2 = 10$, cost is 1.8. Consequently, the workflow completion time is 10 and the total cost workflow is the cost sum for all tasks: $5 + 7 + 4.86 + 1.8 = 18.66$.

In summary, the total cost of workflow of ours is lower than BOSS's and the completion time is equal. This indicates our method can give better resource considering monetary cost and completion time. Because the number of resource is small, the advantage in resource utility will be proved in experiment results.

4 Dynamic Pricing Strategy

While the number of cloud resource providers increases, providers compete against each other to maximize their revenue. So an effective pricing strategy is necessary for providers. In this section, the dynamic pricing strategy is presented. Firstly, two propositions are proposed to prove that dynamic pricing strategy can improve the revenue of providers and decrease the cost of users. Then the process of dynamic pricing strategy is presented.

Proposition 1. Dynamic pricing can improve the revenue of providers.

Proof. We assume provider A has some resources with price p_A, and he has lost several auctions. If he does not change the resource price, it will be low possibility to win successive auctions. If he decreases the price to p'_A, which is still higher than the value of resource, he may be the winner of successive auctions with higher possibility. So it is reasonable for providers to make the price dynamic.

Proposition 2. Dynamic pricing can decrease the cost of users.

Proof. We assume that user B spends the cost c_B on certain resource R when the price of resources is fixed. When the price is dynamic, another resource R$'$ with the same computing ability may decrease the price to c'_B which is lower than c_B. So user B will select resource R$'$ and decrease his cost.

The price of resource in auction changes dynamically according to Formula 1.

$$p_j^{cur'} = \begin{cases} p_j^{cur} & \text{if } j \text{ is winner} \\ p_j^{cur} \cdot \gamma & \text{if } j \text{ is loser and } p_j^{cur} \cdot \gamma > p_j^{min} \\ p_j^{min} & \text{if } j \text{ is loser and } p_j^{cur} \cdot \gamma < p_j^{min} \end{cases} \tag{1}$$

Where p_j^{cur} refers to the current price of resource that provider j owns. p_j^{mini} refers to the bottom price of resource, γ denotes the rate of price reduction. Each provider set peak price, minimum price and change rate of price for resources. When a task starts the auction, providers join the auction and give their bids included computing ability and price. After one auction is over, providers change the resource price according to transaction situation. The process is described as Formula 1.

In conclusion, dynamic pricing is an efficient pricing strategy in cloud market. Providers decrease the price to increase the chance of winning auction and gain more revenue; simultaneously the users will choose the resource with lower price and decrease the cost.

5 Dynamic Pricing-Based Allocation Mechanism (DPAM)

In this section, a novel dynamic pricing-based allocation mechanism is proposed. In auction-based cloud market, the purpose of users is to execute workflows with short completion time and low monetary cost. And the purpose of providers is to sell resources at the best price, so as to gain the highest revenue. In this mechanism, users select the best resource according to the product of completion time and price. And the provider with minimum product will be the winner. After each auction, providers change their resources price according to trade situation. If their competitiveness is weak and lose the auction, they usually decrease price in certain rate to increase competitiveness. Otherwise, if they win, it is effective to keep the price unchanged or increase.

$TC_{i,j}$ is the product of completion time and monetary cost of task i on resource j. (Formula 2). During auction, the task selects the resource with minimum TC as winner.

$$TC_{i,j} = (t_{i,j} + workload_i/ability_j) * (price_j * workload_i/ability_j). \tag{2}$$

For each task, it starts execution only when its parent tasks are finished because of the partial ordering relation of tasks. $t_{i,j}$ refers to the start time of task i executing on resource j which equals to the latest time when its father tasks are finished and the resource j is spare. $workload_i$ refers to workload of task $ability_j$, $price_j$ are computational ability and price per unit time of jth resource respectively. So $workload_i/ability_j$

is the time required for task i executing on resource j. And $(t_{i,j} + workload_i/ability_j)$ refers to the finishing time of task i executing on resource j. $(price_j * workload_i/ability_j)$ is the monetary cost required for task i.

Algorithm. Dynamic pricing-based allocation mechanism
Input: workflows and resources
Output: the schedule of workflows to resources

    1. $tasks \leftarrow [n]$;   /*Assign the tasks to $tasks$ list with specific order*/
    2. $resources \leftarrow [m]$;   /*Assign the resources to $resources$ list */
    3. $i = 1$;
    4.   While $i \leq n$ do
    5. $tempTask \leftarrow ith\ task$;
    6. $j = 1$;
    7.   While $j \leq m$ do
    8. $tempResource \leftarrow jth\ resources$;
    9. $TCs \leftarrow CalculateGoals(tempTask, tempResource)$;
    10.   End
    11. $winner \leftarrow resourceWithMinTC(TCs)$;
               /* select the optimal resource whose TC is minimum */
    12. $tempTask$ pays to $winner$;
    13.   all providers $changePrice$;
    14.   End

There are n tasks and m resources (lines 1–2). Each user starts an auction in order and calculates product of completion time and price for every resource (lines 3–10) and then selects the resource with minimum product (line 11). Then users pay to winner (line 12). At last, all providers change price according to dynamic pricing strategy and join in next auction (line 13).

When workflows are submitted and tasks start auctions, providers give their bids to compete the opportunity for providing services. In the auction, users always select the optimal resource, so the completion time and monetary cost of executing tasks are minimums. In addition, providers change price and increase chances of selling resource, so that they can gain more revenue and higher resource utility.

6 Evaluation

Our mechanism is tested on specific workflow systems and randomly generated workflow systems. In experiments, DPAM and BOSS are evaluated on resource utility and TC.

6.1 Experiment Setup

In specific experiment, to evaluate workflow's balance factor, the workflows are classified into three classes: balanced, semi-balanced and unbalanced [9]. Each workflow has 10 tasks and task workload follows normal distribution $N(1000000, 1000)$.

The speed of seven types of resources is set from 200 to 1200 with an arithmetic progression. The price of resources is setting from the real Amazon Web Services price (http://aws.amazon.com). To implement dynamic pricing strategy, all resources have peak price, bottom price and price change rate. The peak price and bottom price is set according to Amazon Web Services price and the change rate is 20 %. In general experiment, there are 10 workflows and the number of tasks of each workflow increases one by one from 11 to 20. The number of resources increases to 14. Other parameter is same with the first experiment. For the two experiments, the detailed setup can be found in [9].

6.2 Experiment Results

In specific experiment, we use a specific workflow to verify whether DPAM performs better than BOSS on resource utility and TC. In general experiment, we generate 10 workflows with different number of tasks to verify DPAM and BOSS on resource utility and TC.

As shown in Fig. 2, resource utility of DPAM is always higher than that of BOSS. This means that DPAM can improve resource utility compared with BOSS in the three balanced situations. This is because in DPAM the provider changes the resources price with low competitiveness and these resources have more chance to be sold.

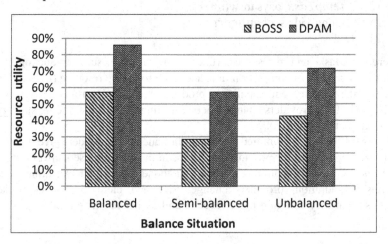

Fig. 2. Comparison of resource utility

From Fig. 3, TCs of DPAM are all lower than those of BOSS. This means that it takes shorter time and lower cost for executing workflows regardless of the balance situations. In DPAM, providers decrease their resource price to improve the competitiveness. So users get the resource with shorter completion time or lower price. So TC becomes smaller.

Figure 4 presents the difference value of resource utility between DPAM and BOSS. This figure shows that difference value is greater than zero regardless of the balanced situation and the number of tasks. This indicates that our mechanism performs

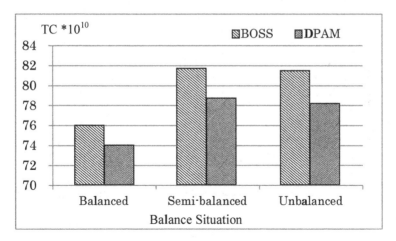

Fig. 3. Comparison of TC

better in resource utility. When the number of tasks and resources increase, resources can be sold with higher possibility because the requirement is greater. Compared with BOSS, more resource is sold by changing price in DPAM especially for resources with low competitiveness that never be sold in BOSS. So DPAM brings higher resource utility than BOSS.

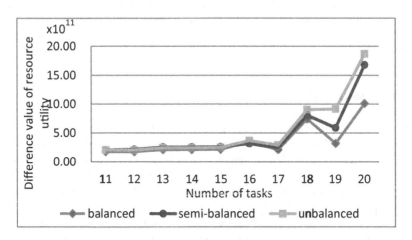

Fig. 4. Difference value of resource utility

Figure 5 describes the difference value of TC between BOSS and DPAM. In this figure, the difference values are no less than zero, which indicates that DPAM brings the shorter completion time and lower monetary cost and the method is effective when the number of tasks increases. In DPAM, resources price is dynamic and the resources with shorter completion time or lower price emerge because of high competitiveness. So TC of DPAM is lower than those of BOSS.

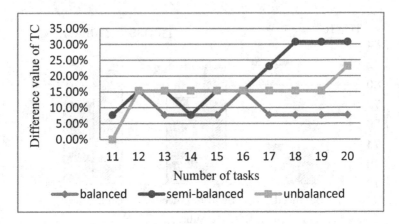

Fig. 5. Difference value of TC

From Figs. 2 and 4, we find the resource utility of DPAM is all higher than that of BOSS. This indicates that the resource can be better utilized when DPAM is applied and the performance of DPAM is still better than BOSS while the number of tasks and resources increases. From Figs. 3 and 5, DPAM performs better than BOSS on TC. And as the number of tasks and resources increase, the advantage is more obvious. This is because the market becomes more competitive and better resources emerge when more providers take part in market. In a word, compared with BOSS, DPAM can improve resource utility and also decrease both monetary cost and completion time.

7 Conclusion and Future Work

In this paper, we propose a dynamic pricing strategy to improve providers' competitiveness in cloud market. And a novel dynamic pricing-based allocation mechanism is presented to allocate resources for workflows. In the mechanism, providers change the price to increase the possibility of selling resources and gain more revenue, which improves resources utility. The users select optimal resources with minimum product, which leads to short completion time and low monetary cost. Finally, we evaluate our proposed mechanism and BOSS. The results show that our mechanism brings high resources utility, short completion time and low monetary cost. However, the dynamic pricing strategy is considered only from auctions of individual provider. Changing one provider's price should refer other related providers' auctions. In addition, how to set initial price and the price change ratio is also an interesting research filed.

Acknowledgement. This work is partially supported by Australian Research Council Linkage Projects LP0990393 and LP130100324, and Chinese National Natural Science Foundation Project NO 61300169 and the key teacher training project of Anhui University.

References

1. Foster, I., Yong, Z., Raicu, I.: Cloud computing and grid computing 360-degree compared. In: Grid Computing Environments Workshop, 2008. GCE 2008, pp. 1–10 (2008)
2. Juve, G., Deelman, E.: Scientific workflows and clouds. J. Crossroads **16**, 14–18 (2010)
3. Cui, L., Shiyong, L.: Scheduling scientific workflows elastically for cloud computing. In: 2011 IEEE International Conference on Cloud Computing (CLOUD), pp. 746–747 (2011)
4. Tram Truong, H.T. Chen-Khong, T.: An auction-based resource allocation model for green cloud computing. In: 2013 IEEE International Conference on Cloud Engineering (IC2E), pp. 269–278 (2013)
5. Vinu Prasad, G., Rao, S., Prasad, A.: A combinatorial auction mechanism for multiple resource procurement in cloud computing. In: 2012 12th International Conference on Intelligent Systems Design and Applications (ISDA), pp. 337–344 (2012)
6. Rahmanl, M.A., Rahman, R.M.: CAPMAuction: reputation indexed auction model for resource allocation in grid computing. In: 2012 7th International Conference on Electrical and Computer Engineering (2012)
7. Qu, H., Ryzhov, I.O., Fu, M.C.: Learning logistic demand curves in business-to-business pricing. In: Proceedings of the 2013 Winter Simulation Conference: Simulation: Making Decisions in a Complex World, pp. 29–40. IEEE Press (2013)
8. Prasad, A.S., Rao, S.: A mechanism design approach to resource procurement in cloud computing. J. IEEE Trans. Compt. **63**, 17–30 (2014)
9. Fard, H.M., Prodan, R., Fahringer, T.: A truthful dynamic workflow scheduling mechanism for commercial multicloud environments. J. IEEE Trans. Parallel Distrib. Syst. **24**, 1203–1212 (2013)
10. Wood, T., Shenoy, P.J., Venkataramani, A.: Black-box and gray-box strategies for virtual machine migration. In: NSDI, p. 17 (2007)
11. Gorlach, K.. Leymann, F.: Dynamic service provisioning for the cloud. In: 2012 IEEE Ninth International Conference on Services Computing (SCC), pp. 555–561 (2012)
12. Shi, X., Zhao, Y.: Dynamic resource scheduling and workflow management in cloud computing. In: Web Information Systems Engineering - Wise 2010 Workshops, pp. 440–448 (2011)
13. Ludwig, S.A.: Particle swarm optimization approach with parameter-wise hill-climbing heuristic for task allocation of workflow applications on the cloud. In: 2013 IEEE 25th International Conference on Tools with Artificial Intelligence (ICTAI), pp. 201–206 (2013)
14. Sharma, B., Thulasiram, R.K., Thulasiraman, P.: Pricing cloud compute commodities: a novel financial economic model. In: 2012 12th IEEE/ACM International Symposium on Cluster, Cloud and Grid Computing (CCGrid), pp. 451–457 (2012)
15. Niu, D., Feng, C., Li, B.: Pricing cloud bandwidth reservations under demand uncertainty. In: ACM SIGMETRICS Performance Evaluation Review, pp. 151–162. ACM (2012)
16. Xu, H., Li, B.: Maximizing revenue with dynamic cloud pricing: the infinite horizon case. In: 2012 IEEE International Conference on Communications (ICC), pp. 2929–2933. IEEE (2012)
17. Teng, F., Magoules, F.: Resource pricing and equilibrium allocation policy in cloud computing. In: 2010 IEEE 10th International Conference on Computer and Information Technology (CIT), pp. 195–202. IEEE (2010)
18. Mihailescu, M., Teo, Y.M.: On economic and computational-efficient resource pricing in large distributed systems. In: 2010 10th IEEE/ACM International Conference on Cluster, Cloud and Grid Computing (CCGrid), pp. 838–843. IEEE (2010)

On-the-Fly Performance-Aware Human Resource Allocation in the Business Process Management Systems Environment Using Naïve Bayes

Arif Wibisono[1], Amna Shifia Nisafani[1(✉)],
Hyerim Bae[2], and You-Jin Park[3]

[1] Information Systems Department,
Institut Teknologi Sepuluh Nopember, Jalan Raya ITS Sukolilo,
Surabaya, Indonesia
{wibisono,amna}@is.its.ac.id
[2] Industrial Engineering Department, Pusan National University,
Busandaehak-ro 63 beon-gil, Geumjeong-gu, Busan
Republic of Korea
hrbae@pusan.ac.kr
[3] School of Business Administration, Chung-Ang University,
84 Heukseok-dong, Dongjak-gu, Seoul, Republic of Korea
eugenepark@cau.ac.kr

Abstract. Traditionally, resource allocation problem has been considered as one of the important issues in business process management to maintain the acceptable level of each activity completion time which can reduce the total completion time. Especially, the complexity of managing resources increases when the resource type is human because performance of each human resource might fluctuate over time due to various unpredicted factors. Hence, upfront planning of the resource allocation might be unsuitable in this matter. Therefore, this study proposes an on-the-fly resource allocation using Naïve Bayes to manage human resources more efficiently. The term on-the-fly here indicates that the resource allocation planning will be frequently updated and executed during the execution time by considering recent human resource performances. In this paper, we will show the proposed approach exceeds other resource allocation approaches in terms of total completion time.

Keywords: On-the-fly resource allocation · Machine learning · Dispatching rules · Resource-based priority rules

1 Introduction

Recent business competitiveness requires every company to remain efficient in order to survive in highly complex business environments. Therefore, more and more softwares related to business process management systems (BPMS) are deployed in many

© Springer International Publishing Switzerland 2015
J. Bae et al. (Eds.): AP-BPM 2015, LNBIP 219, pp. 70–80, 2015.
DOI: 10.1007/978-3-319-19509-4_6

companies. Here, BPMS focuses on the planning, execution, control, monitoring, and evaluation of business process (BP) execution to obtain some important efficiency extents. For this reason, several scheduling approaches were introduced to help BPMS to better organize the resources involved in the BP execution [1–9].

In general, there are two types of resources in BP execution: human and machine [8]. When machine is prevalent in many manufacturing process, human is dominant in numerous organizational processes such as order-to-cash, quote-to-order, procure-to-pay, issue-to-resolution, and application to approval [10]. In terms of resource allocation, machine-related processes are easier to maintain due to the lower variability of the machine performances. In contrast, human performances are oscillating continuously due to the differences of knowledge and physical/emotional conditions. Thus, it is almost impossible to ask human to work in a regular pace during his/her daily work time. To illustrate, a worker might work industriously in his/her first three hours and then the performance systematically declines around the lunch time. After doing lunch, the performance increases; however, it is not as the same pace as in the early morning, and it is getting steady from 3 PM to the end of the work time.

The issue of business process scheduling has received considerable attention among researchers. Zhao and Stohr developed method to reduce the amount of rework in claim handling system [7]. Bae *et al.* [1] proposed a methodology using mixed integer programing (MIP) for BP execution plan by taking into consideration business process semantics and alternative path in the business process management structure. Eder *et al.* [2] built personal schedule to forecast future incoming jobs in which organizations can decrease both the turnaround time and the rate of time-constraint violation.

One of the limitations with the most recent papers in business process scheduling is that they focus on the upfront planning (see Sect. 2.1 Literature Reviews). Here, upfront planning means that the resource allocation planning has been completely established before the execution takes place. The upfront planning might be unable to fully accommodate resource performance dynamics during execution. Hence, it might fail to allocate right resource(s) in doing an incoming job.

This study aims to develop an on-the-fly performance-aware resource allocation by incorporating Naïve Bayes in the proposed algorithm. To this point, the resource allocation planning will be rigorously updated and (then) executed given the evident of each human resource performances. By doing so, we can enhance the resource performance prediction, thus making a better resource allocation. For this reason, we expect a better human resource allocation in terms of the completion time. Most of the work of this paper including the algorithm basis and the real-world study case is an extension of Nisafani *et al.* [8]. The results of this study prove to be very useful for any process efficiency oriented managers, especially those who are responsible for managing human-intensive processes. The reminder sections of this paper are organized as follows: Sect. 2 defines the literature reviews, Sect. 3 explains our proposed approach, Sect. 4 provides the experiments of our model, and finally, Sect. 5 presents the conclusions of the research.

2   Literature Reviews

2.1   Resource Allocation in the Business Process Management Systems Environment

Most available methods of the resource allocation in business process have been focusing on the upfront resource allocations planning [11]. Wu *et al.* [6] predicted the future resource behavior using workload dynamics. Huang *et al.* [3] proposed a resource allocation algorithm that utilizes Markov decision process and solved using reinforcement learning. Ha *et al.* [12] introduced process execution rule to fairly distribute workload for each involved resource (which called as agent). Huang *et al.* [4] suggested a method of resource behavior measurement using four crucial factors to improve BP execution namely availability, corporation, preference, and competence.

Most of the studies reviewed so far; however, suffers from the fact that upfront resource planning can be unsuitable to capture the resource performance dynamics (especially whenever the resource type is human). Most of the studies assume that resource performances are uniformly distributed during time horizon, in fact they are not. Here, the forecast accuracy of the resource performances declines heavily as the forecast horizon increases [13]. Hence, it is necessary to consider a run-time oriented and performance-aware resource allocation in business process for improving the system performance. There is a limited study that starts investigating effective methods for on-the-fly performance-aware resource allocation. Nisafani *et al.* [8] simulated the on-the-fly performance-aware resource allocation on a real-world semi-automatic business process and recommended resource allocation algorithm to use Bayesian network (BN) as a model. The BN incorporates several factors in BP execution such as workload, inter-arrival time, daytime, and working hours [8]. The result showed that the existing resource allocation exceeded four resource-based priority rules namely index-ordered, shortest-idled, longest-idled, and random allocation in terms of average completion time, average waiting time, and average cycle time [8].

The weakness of Nisafani *et al.*'s approach is that the employed heuristic-based BN introduces a BN structure which quality has never been measured statistically. At this point, the BN designs the relationship dependencies among factors involved in BP execution such as resource performance, queue, and inter-arrival time. In addition, there is no exact formula to model BN among aforementioned factors thus heuristic approach is used. That's why expert is assumed has a comprehensive understanding about the intertwined factors. Unfortunately, the understanding might be mistaken. For instance, an individual with a highly imposed workload does not basically demonstrate performance decline. In contrast, an individual even though is imposed with long queues does not essentially increase his/her performance. As a result, a statistical measurement is required to appraise the BN structure. Surely, a better BN structure will introduce a better prediction.

It is known that measuring BN structure does not automatically produce a good BN structure, rather, additional time consuming algorithm (such as K2 Algorithm) should be performed later to construct a BN structure. To increase the prediction accuracy in the long run, a periodic invocation of the BN constructing algorithm is also necessary and, consequently, finding a method that averts extra computing time to establish a new

BN structure or reestablish existing BN structure as well as the expert's erroneous BN causal relationship making is indispensable. Thus, we employ Naïve Bayes Assumption (NBA) to formulate our proposed approach. In the NBA, each factor is considered independently to others except to one factor (which we called as "target"). Further, even though in the long run, we find that two or more factors are dependent each other, the NBA still demonstrate a good conjecture [14]. By obtaining this characteristic, we can discourage any unnecessary algorithm and erroneous expert judgment in making BN structure while still maintaining prediction accuracy.

2.2 Bayesian Network and Naïve Bayes

A Bayesian Network (BN) which consists of a set of nodes and links is a causal representation model and is useful to model uncertainty [15]. A BN assumes a form of Directed Acyclic Graph (DAG) in which every node within BN (we called as BN variables) denotes random variables and every link within BN characterizes probabilistic dependences of BN variables [16]. These relationships are then measured by associating a conditional probability table with each BN variable. Usually, let $G = (V,E)$ be a DAG with a node set V and a link set E, and let $X = (X_v)_{v \in V}$ be a set of random variables indexed by v.

Naïve Bayes is a subset of BN. It has a simple structure with one target node as the parent node of all other nodes and it has a restriction that other structure to occur [17]. The benefit of employing Naïve Bayes is that it reduces the complex calculation efforts compared to general BN due to its simple structure (see Fig. 1). That is, it is possible to avoid a more complex calculation because Naïve Bayes supposes that each node is independent to other nodes except to the target node. The independent assumption might be problematic [17], however Langley et al. [18] have found that Naïve Bayes has surpassed other complex algorithms for a problem with highly large datasets, especially when factors (in which each factor is represented as an attribute in data mining terminology) are independent of each other.

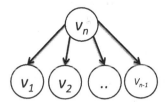

Fig. 1. A simple Naïve Bayes structure

3 Proposed Approach

3.1 Process Structure

Definition 1 (Process Structure). The definition of process structure is adapted from [5]. A process structure is a directed graph $P = (A, L, F)$ consisting of sets of node A, sets of arcs L and the labeling function F

- $A = \{a_i | i = 1, \ldots, N\}$ is the set of activities, where a_i is the ith activity and N is the total number of activities in P.
- $F \subseteq \{(f_s, f_m)\}$ is the set of labeling function, where f_s is the split function and f_m is the merge function
- $L \subseteq \{(a_{i-}, a_{j+}) \mid a_{i-}, a_{i+} \in A$ and $i+ \neq i-\}$ is the set of links where an element (a_{i-}, a_{i+}) represents a_{i-} immediately precedes, a_{j+}.
- For a split activity a_j, such that $|SA_i| > 1$, where $SA_i = \{a_{j+} |(a_i, a_{i+}) \in L\}$, $f(a_i) =$ "AND" if all a_{i+} 's should be executed; otherwise $f(a_{i+}) =$ 'OR'.
- For a merger activity a_j such that $|MA_i| > 1$, where $MA = \{a_{j-} |(a_{i-}, a_j) \in L\}$,
- $f(a_{j-}) =$ 'AND' if all a_i should be executed; otherwise $f(a_{j-}) =$ 'OR'.
- For a merger activity a_j such that $|P| > 1$, where $P = \{a_i |(a_i, a_j) \in L\}$, $f(a_j) =$ 'AND' if all a_i should be executed; otherwise $f(a_i) =$ 'OR'.

3.2 Naïve Bayes in the Proposed Approach

We denote the Naïve Bayes incorporated in our algorithm as Naïve Bayes Model (NBM). The NBM consists of five nodes: Human Performance, Activity, Queue, Inter-arrival and Day Time (see Fig. 2). Each node represents factors involved in BP Execution. There are two types of node: target node and child node.

The target node is the factor to characterize human performance while the four other factors are the child nodes. Here, the human performance is something to predict given information from all child nodes. The detail description of each node can be seen in Table 1.

Table 1. Nodes in the NBM

No	Nodes/Factors	Possible states	Notes
1	Human performance	Low, Medium, High	Human resource performance prediction (**Target Node**)
2	Queue	Low, Medium, High	Queue in front of the activity
3	Inter-arrival rate	Short, Medium, Long	Average of the systems' inter-arrival time/hours.
4	Performer	{human resource name}	
5	Activity	{activity name}	
6	Day time	Morning, Afternoon, Evening	The working shift

To illustrate (see Fig. 3), during simulation at time t, we need to determine the best performer (human resource) to carry out a job in the activity a_i (let say "For request"). Suppose, we observe that current situations at time t are: the daytime is in the *morning* (*first shift*), the inter-arrival rate is *short*, and the imposed queue is *low*. Hence, from Fig. 3, we can see that we should select p_3 because it introduces probability value of 80 %. The second alternative whenever p_3 is unavailable is p_1 because p_1 values 15 %. Here 80 % and 15 % are the possibilities that p_3 and p_1 will have higher human performances.

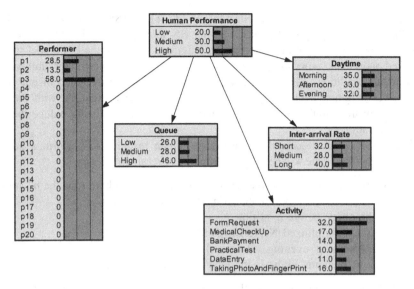

Fig. 2. Naïve Bayes Model (NBM)

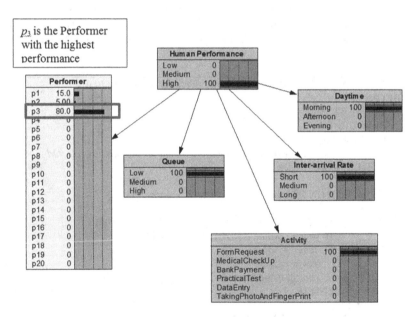

Fig. 3. Example of NBM with some evident at time = t, Selected performer will be determined by the probability of a performer to produce a higher human performance.

3.3 Naïve Bayes Selection Rule (NBSR) Algorithm

We propose *NBSR* (t, a_i) as on-the-fly performance-aware algorithm that uses NBM described in the earlier section. NBSR is an extension of BSR which is an allocation resource algorithm proposed by Nisafani *et al.* [8]. NBSR is similar with BSR algorithm except employing the NBM as the BN model. Nisafani *et al.* [8] accommodates some previously determined expert judgment factors in the BN model such as perceived workload, working pressure, technology support, performer ability, and environment condition. All of the aforementioned factors are excluded in our NBM because the judgment factors are static and might not be compatible with the randomization in the simulation software in the long run; hence, it will reduce the prediction accuracy.

The NBSR is to allocate the appropriate human resource to perform a process instance in the a_i at time t. NBSR (t, a_i) employs several parameters:

- $R_a = \{r_n | n = 1, 2, \ldots, N\}$ is the set of human resources where r_n is the nth human resource and N is the total number of resources employed in a_i
- $Q_a(t)$ is the queue before a_i at time t
- BN represents the utilized Naïve Bayes Model
- $D_a(t)$ \in {morning, evening, afternoon} is daytime at time t
- $I(t)$ \in {low, medium, high} is the inter-arrival rate at time t

Figure 4 denotes the NBSR. Here, the algorithm forecast the performance of human resources and assigns an incoming job to human resource with the highest performance predicted from the NBM. Each resource will be recognized with non-negative and unique index, and the NBSR will select the index with the possibility to introduce a higher performance. Also, one of the algorithm component is to invoke Naïve Bayes

```
1    FUNCTION SELECT RESOURCE (aᵢ, Qₐ(t), Rₐ, BN, Dₐ (t), I(t))
2    BEGIN
3        BOOLEAN loop := TRUE ;
4        RESOURCE res ;
5        DOUBLE temp := -9999;
6            //very big negative number, indicating no human resource is selected
7        WHILE (loop = TRUE)
8        {
9                FOR (INT index :=0; index<size(Rₐ) ; index++)
10               {
11                    value : = do_inference(aᵢ, Qₐ(t), Rₐ, BN, Dₐ (t), I(t));
12                    IF (temp < value && rᵢₙdₑₓ IS IDLE) THEN
13                        temp := value;
14                        res = rᵢₙdₑₓ ;
15                        // rᵢₙdₑₓ is the resource in the Rₐ with index = index
16                    END IF
17               }
18               IF (res != NULL) THEN
19                    loop := FALSE;
20               END IF
21       }
22        RETURN res;
22    END
```

Fig. 4. NBSR Algorithm

Model (see Function *do_inference* in line 11). The *do_inference* is defined as a probability function in BN as follows:

$P(Human\_Performance = "High" | Activity = a_i, Queue = Q_a(t), Humanresource = r_n, Daytime = D_a(t), Inter\text{-}arrival = I(t)).$

By using this function, we select the human resource given activity a_i, Queue $Q_a(t)$, inter-arrival $I(t)$, human resource r_n, and Day time $D_a(t)$.

4   Experiments and Results

4.1   Static Rules for Resource Allocation

Below selection rules (see Table 2) are adapted from a simulation book which can be applied in the manufacturing process for managing machine resources [19]. Since, our approach is to manage human resource, comparing these static rules with our proposed approach is relevant because their ability to measure our algorithm in accommodating the human performance dynamics.

Table 2. Static rules for resource allocation [19]

No	Priority rule	Description
1	ORDER	Select from the free resources in the preferred order
2	LIDDLE	Select the resource that has the largest idle to date
3	SIDDLE	Select the resource that has the smallest idle to date
4	RANDOM	Select randomly among all free resources

4.2   Experiment Results

This study uses a real world semi-automatic business process mentioned in Nisafani *et al.* [8]. The process is the driver lisence application process conducted in Indonesia. The business process consists of 8 activities in which 6 of them were performed by the assigned police officers (see Fig. 5). There is no officer assigned for two activities (theoretical test and practical test) since both activities are conducted by the applicants.

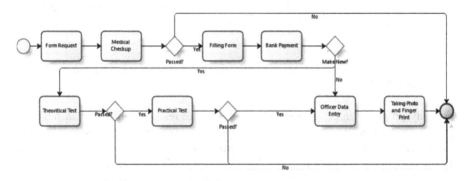

Fig. 5. Driver license application process

Every officer is responsible to his/her activity and officer transfer among activities is not allowed. The detail parameters of the simulated system (such as human resource processing time, interarrival time, etc.) is available at Nisafani *et al.* [8]. In general the simulation running time is 13 hours per day and is consisting of three working shifts: Morning (8 AM–12 AM), Afternoon (12 AM–4 AM), and Evening (4 PM–9 PM). Nisafani *et al.* [8] recorded each resource performance within three shifts, from which human performance distribution was developed. Most of the human performance distribution followed normal distribution. In addition, the replication number is 10 and the average instance numbers per replication is 1500.

We compare NBSR with the static rules described in Table 2 and BSR. The comparison is available in Table 3. In general, NBSR outperformed all static rules and BSR in terms of the mean and standard deviation. Hence, we can say that NBSR can accomodate the human performance fluctuations. However, even though NBSR demostrates a better completion time than BSR, the difference between the NBSR and BSR are very near. We suspect that the number of the human resources in each activity (three officers) and the simulation duration are responsible to the small distance of the completion time between the BSR and NBSR. A larger number of the human resources per activity as well as a longer simulation duration might help us to clearly understand how the behavior of the NBSR and BSR when the system escalates.

Table 3. Experiments result in terms of the average completion time (Bold Numbers indicates the Lowest Completion Time for each replication)

Replication#	RANDOM	ORDER	SIDDLE	LIDDLE	BSR	NBSR
1	1096.167	1567.904	2590.262	1808.748	922.6673	**921.3511**
2	2232.817	3127.327	2016.082	2088.21	1364.995	**1362.517**
3	1171.825	1945.549	2502.724	1656.001	**828.006**	826.453
4	983.3494	407.1013	2421.051	1953.586	919.0942	**917.6344**
5	2327.259	1139.865	1789.55	1674.413	1092.204	**1088.304**
6	1403.965	**1181.13**	3668.057	779.2716	1954.795	1948.438
7	1034.707	**1195.526**	1702.11	3018.147	1547.087	1560.595
8	638.2284	1597.009	952.9931	664.4587	912.9886	**896.3806**
9	1842.827	1664.555	**1363.279**	3165.451	1670.024	1666.54
10	925.5398	1544.924	1514.428	1987.32	1326.165	**1321.235**
Mean	1365.668	1537.089	2052.054	1879.561	1253.803	**1250.945**
Standard Deviation	547.2241	662.916	735.1494	761.2297	362.4221	**361.4053**

5 Conclusion

This study proposes an on-the-fly performance-aware resource allocation in business process management. We utilize Naïve Bayes Model in the Naïve Bayes Selection Rule (NBSR) algorithm for selecting the best performer to accomplish an incoming task. We compare our approach with four static rules and previously developed BSR. The result shows that NBSR surpasses all the aforementioned rules. Therefore, the result indicates that Naïve Bayes approach is beneficial to model complex relationships among factors in the BP execution. Future research might accomodate resource transfer among activities and incorporate workflow resource patterns [20] in the business process management. It is necessary to conduct a longer simulation time to thoroughly observe how the NBSR works in the long run.

References

1. Bae, H., Lee, S., Moon, I.M.: Planning of business process execution in business process management environments. Inf. Sci. **268**, 357–369 (2014)
2. Eder, J., Pichler, H., Gruber, W., Ninaus, M.: Personal schedules for workflow systems. In: van der Aalst, W.M., ter Hofstede, A.H., Weske, M. (eds.) BPM 2003. LNCS, vol. 2678, pp. 216–231. Springer, Heidelberg (2003)
3. Huang, Z., van der Aalst, W.M.P., Lu, X., Duan, H.: Reinforcement learning based resource allocation in business process management. Data Knowl. Eng. **70**(1), 127–145 (2011)
4. Huang, Z., Lu, X., Duan, H.: Resource behaviour measure and application in business process management. Expert Syst. Appl. **39**, 6458–6468 (2012)
5. Rhee, S.-H., Bae, H., Kim, Y.: A dispatching rule for efficient workflow. Concurr. Eng. Res. Appl. **12**(4), 305–318 (2004)
6. Wu, B.Y., Chi, C.-H., Chen, Z., Gu, M., Sun, J.G.: Workflow-based resource allocation to optimize overall performance of composite services. Future Gener. Comput. Syst. **25**, 199–212 (2009)
7. Zhao, J.L., Stohr, E.A.: Temporal workflow management in a claim handling system. In: The International Joint Conference on Work Activities Coordination and Collaboration, New York, pp. 187–195 (1999)
8. Nisafani, A.S., Wibisono, A., Kim, S., Bae, H.: Bayesian selection rule for human-resource selection in business process management systems. J. Soc. e-Business Stud. **17**(1) (2014
9. Yahya, B., Wibisono, A., Bae, H., Ryu, K.: Bayesian network for finding best in BPM Environment. ICIC Express Lett. **2**(2), 473–479 (2011)
10. Dumas, M., La Rosa, M., Mendling, J., Reijers, H.: Fundamentals of Business Process Management. Springer, Berlin (2013)
11. Wibisono, A., Nisafani, A.S., Bae, H., Park, Y.-J.: Early detection for supply chain disruption using bayesian network. In: Asia Pacific BPM, Brisbane, Australia (2014)
12. Ha, B.-H., Bae, J., Park, Y.T.: Development of process execution rules for workload balancing on agents. Data Knowl. Eng. **56**, 64–84 (2006)
13. Armstrong, J.: Principles of Forecasting: A Handbook for Researcher and Practitioners. Springer, New York (2001)
14. Whitten, I., Frank, E., Hall, M.: Data Mining: Practical Machine Learning Tools and Techniques: Practical Learning Tools and Techniques, 3rd edn. Morgan Kauffman, San Francisco (2011)
15. Kao, H.-Y., Huang, C.-H., Li, H.-L.: Supply chain diagnostics with dynamic bayesian networks. Comput. Ind. Eng. **49**(2), 339–347 (2005)
16. Gregory, C., Herskovits, E.: A Bayesian method for the induction of probabilistics network from data. Mach. Learn. **9**(4), 309–347 (1992)
17. Cheng, J., Grainer, R.: Comparing Bayesian Network Classifiers, pp. 101–108. Morgan Kauffman Publishers, San Francisco (1999)

18. Langley, P., Iba, W.I., Thompson, K.: An analysis of Bayesian classifiers. In: Tenth National Conference of Artifical Intelligence (1992)
19. Pritsker, A., O'Reilly, J.: Simulation with Visual SLAM and AweSim. System Publishing Corporation, West Lafayette (1999)
20. Russel, N., ter Hofstede, A., Edmond, D., van der Aalst, W.: Workflow Resource Pattern (2004)

Algorithms for Process Analysis

Run JTA in JTang: Modeling in Artifact-Centric Model and Running in Activity-Centric Environment

Binbin Fan, Ying Li$^{(\boxtimes)}$, Shengpeng Liu, and Yangtao Zhang

College of Computer Science, Zhejiang University, Hangzhou, China
cnliying@zju.edu.cn

Abstract. Nowadays, there are two major kinds of business process models: activity-centric model and artifact-centric model. Activity-centric model is widely used in industry while artifact-centric model has few engines to support its execution, which limits its popularization. One efficient way to address this problem is to transform artifact-centric model to activity-centric model directly so the transformed model can be published in activity-centric engine. However, the transformation is not easy as we need to handle differences between models. In this paper, JTangFlow model and JTAFlow model are defined as activity-centric model and artifact-centric model respectively. A transforming algorithm is proposed in level of schema. In case study, A JTAFlow model is designed in its designer, the xml file is transformed automatically and run in JTangFlow engine well.

Keywords: Business process modeling · Activity-centric · Artifact-centric · Model transformation

1 Introduction

Workflow modeling has received much attention since 1990s, because it's a bridge connecting business manager and computer system. A well defined workflow model makes a good workflow system when analysis, improvement, control, management, and enactment of the processes. So, research on workflow modeling is always developed. Today, there are two major kinds of workflow models: activity-centric model and artifact-centric model. Activity-centric model language which depending on petri-net was put forward early in 1990s, developed in these decades. BPEL is an executable business process modeling language and acts as current standard [1]. YAWL aims to support most of the workflow patterns [2,3]. Nowadays, activity-centric model is widely accepted, such as BPMN. Based on BPMN, Process engines such as JBPM, Activiti [4], Bonita [5] and JTangFlow [6] are able to enact the control flow of a process by a given process model. Other popular workflow systems include Taverna, Kepler, Triana and so on.

© Springer International Publishing Switzerland 2015
J. Bae et al. (Eds.): AP-BPM 2015, LNBIP 219, pp. 83–97, 2015.
DOI: 10.1007/978-3-319-19509-4_7

Artifact-centric business process modeling approach appears natural to business managers and software engineers. This idea was first articulated by Nigam and Caswell in their seminal paper [7]. The artifact-centric BP models make a significant step towards complete specification of BP semantics at the logical level.There are many artifact-centric business process modeling methods now, such as EZFlow [8], PHILharmonicFlows [9], GSM-based business artifacts [10]. However, there is few software system based on artifact-centric BP modeling. The lack of Artifact-centric BP modeling engine leads to a gap, which limits its practise application.

In this paper, we try to eliminate this gap. The main idea of this paper is to demonstrate that it is possible to automatically translate artifact-centric BPs to JTangFlow models and to build an application to execute the BPs on the JTangFlow engine. Figure 1 shows the how the translation work. With an example from this application,we illustrate a mapping approach between artifact-centric model and JTangFlow specifications and then into executable processes. People use Artifact-centric Desginer, publish models in Activity-centric Engine and the published models perform normally as Activity-centric systems. We are optimistic that automated translation of business processes is an achievable goal, but there are many technical problems to be solved.

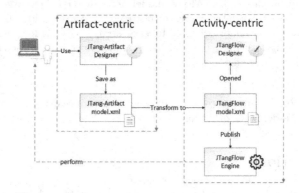

Fig. 1. How the translation work

The remainder of the paper is organized as follows. Section 2 outlines the formal defination of JTangFlow model and JTang-Artifact model. In Sect. 3, an algorithm is put forward to map JTang-Artifact model to JTangFlow model in level of schema. Section 4 evaluates the execution performance on the example and discusses several technical issues associated with the translation. Related work is introduced in Sect. 5. A brief conclusion is included in Sect. 6.

2   JTangFlow and JTAFlow

In order to address the technical problems in mapping a artifact-oriented process to an activity-oriented process, representative and concrete models (for both processes) are needed. In this section, two models are introduced, namely JTangFlow and JTAFlow(short for JTang-Artifact Flow). The reason that we choose these two models is, they are the models at level of execution. JTangFlow is the basic model of JTangFlow system, which is an excellent workflow design and runtime environment engine from JTangSoftware. It has been worked perfectly for many BPEL-like business model cases (http://www.jtang.cn/?page_id=321). By mapping JTAFlow to JTangFlow, we can get a guaranteed performance.

2.1 JTangFlow

JTangFlow is our flow-oriented process engine with solid theoretical formalization and we has all control of the source of it to do experiments for both the formal mapping and implementation. In this sub-section, we introduce the JTangFlow enactment model in aspect of schema, instance and snapshot.

Definition 2-1. A JTangFlow Schema is a tuple(A, D, C, R, E, P, M_{DA}, M_{EA}, M_{PA}), where

1. A is a set of activities, the type of activity $Type_A \in$ {Manual, Auto, System, SubPro, Start, End, Choice, Syncro, Vote};
2. D = {(Name ,$Type_D$, InitialValue) | $Type_D \in$ {basic, userdef, array}, Name is a string, InitialValue is a string};
3. C is a set of conditions, which are expressions involving attributes in D;
4. R = {(r_1, r_2)}, $r_1 \in$ {Sequence, Jion, Split}, $r_2 \in$ {AND, OR, XOR};
5. E is a set of events, which consist of (activity $\in$ A, condition $\in$ C, routemode $\in$ R), meaning that when the activity is finished, the condition is satisfied and the routemode is admitted, the event will be triggered;
6. P is a set of participants consist of Users, Roles and Organizations;
7. M_{DA} is a set of (data $\in$ D, activity $\in$ A), binding attributes to activities;
8. M_{EA} is a set of (event $\in$ E, activity $\in$ A), binding events to activities;
9. M_{PA} is a set of (expression involving participants in P, activity $\in$ A), binding participants to activities;

Example 2-1. Figure 2 shows a JTangFlow Schema example of student business trip application process. Each box represents an activity and the trapezoid one stands for system-activity (This activity does nothing). Each activity maps to many elements including data (Application, Teacher opinion, Director opinion), participants (student, teacher, director), events and so on. According to conditions and route mode, activities connect each other by events. When a student want to apply a trip, the process will be like this. Firstly, the student submits an application. Then, his teacher checks this application and chooses to admit or reject it. Thirdly, if the teacher admits the application, the director checks the application and if he admit it, the application is accepted. Else, the application is rejected. Meanwhile, the process is done.

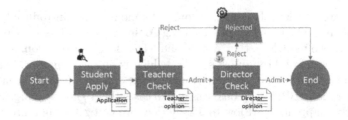

Fig. 2. JTangFlow Schema example: student business trip application process schema

Definition 2-2. A JTangFlow Instance is a tuple(PI, AI, W, DI), where

1. PI is a set of process instances, each instance is a tuple(PIid, Pid, PIName, Starterid), PIid is an unique process instance ID, Pid is process ID, PIName is name of process instance, Starterid is id of process instance starter;
2. AI is a set of activity instances, each instance is a triple(AIid, Aid, PIid, AState), AIid is activity instance ID, Aid is activity ID, PIid acts as foreign key, AState is state of activity such as ready, running, pause, terminate, finished, cancel, working;
3. W is a set of workitems, each workitem is a tuple(Wid, AIid, Participant, WState), Wid is workitem ID, AIid acts as foreign key, participant is participant ID, WState is state of workitem such as ready, running, pause, terminate, finished, cancel, working;
4. DI is a set of data instances, each instance is a tuple(DIid, Did, PIid, Value), DIid is data instance ID, Did is data ID, PIid acts as foreign key, Value is the value of data instance.

A JTangFlow Instance is given in Example 2-3: JTangFlow Snapshot example.

Definition 2-3. A JTangFlow system snapshot is a tuple (Instances, Schemas), where Instances is a set of JTangFlow instances, Schemas is a set of JTangFlow schemas. Both of them are defined before. Figure 3 shows JTangFlow snapshot

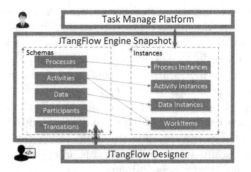

Fig. 3. JTangFlow Snapshot Definition

Table 1. ProcessInstance

PIid	Pid	PIName	Starterid
1	Application	Application1	Bob
2	Application	Application2	Alice

Table 2. ActivityInstance

AIid	Aid	PIid	AState
1	Start	1	finished
2	StudentApply	1	running
3	Start	2	finished
4	StudentApply	2	finished
5	TeacherCheck	2	running

Table 3. Workitem

Wid	AIid	Participant	WState
1	2	Bob	ready
2	4	Alice	finished
3	5	DoctorA	ready

Table 4. DataInstance

DIid	Did	PIid	Value
1	Application	1	2015/1/30;Bob
2	Application	2	2015/1/30;Alice
3	TeacherOpinion	2	false

definition, which tells us how instances and schemas work together. After published, schemas are stored in relational tables of engine database, which the engine can recognise. Each instance binds its schema by recording schema ID.

Example 2-3. Tables 1, 2, 3 and 4 show a JTangFlow Snapshot example based on the JTangFlow schema example in Fig. 2. Table 1 records 2 process instances: Application1 started by Bob and Application2 started by Alice. Activity instances are stored in Table 2, which tell us process instance 1(Application1) is in stage of StudentApply, 2(Application2) is in stage of TeacherCheck. Table 3 shows activity instance 2(StudentApply) is ready for Bob, activity instance 4(StudentApply) is finished and activity instance 5(TeacherCheck) is ready for DoctorA. Table 4 records data instances. Value of data instance TeacherOpinion is initially false as the activity TeacherCheck has not been done. By using PIid, activity instances are connected with process instances. Workitems use AIid to connect with activity instances. And data instances use PIid because they belong to process instances. In this way, these tables work together and support execution of process instances.

2.2 JTAFlow

JTAFlow is short for JTang-Artifact Flow. It is an artifact-centric model which includes Entities and States. In this sub-section, we introduce the JTAFlow model in aspect of schema, instance and snapshot respectively.

Definition 2-4. A JTAFlow Schema is a tuple(En, T, S, C, R, Ev, P, M_{EnTS}, M_{PTS}, M_{EvTS}), where

1. En is an entity tree involving all data(business data and control data) related to workflow, each branch has its path, each child has its path and data attribute;

2. T is a set of tasks, the type of task $Type_T \in$ {Manual, Auto, System, SubPro, Start, End, Choice, Syncro, Vote};
3. S is a set of states, consist of a set of attributes which represent the state of these attributes; in fact, S is snapshot of subset of entity;
4. C is a set of conditions, which are expressions involving attributes in En;
5. $R = \{(r_1, r_2)\}$, $r_1 \in$ {Sequence, Jion, Split}, $r_2 \in$ {AND, OR, XOR};
6. Ev is a set of events, which consist of (task $\in$ T $\bigcup$ state $\in$ S, condition $\in$ C, routemode $\in$ R), meaning that when the task is finished or the state is achieved, the condition is satisfied and the routemode is admitted, the event will be triggered;
7. P is a set of participants consist of Users, Roles and Organizations;
8. M_{EnTS} is a set of (attribute $\in$ En, taskstate $\in$ T $\bigcup$ S), binding attributes to tasks and states;
9. M_{EvTS} is a set of (event $\in$ Ev, taskstate $\in$ T $\bigcup$ S), binding events to tasks and states;
10. M_{PTS} is a set of (expression involving participants in P, taskstate $\in$ T $\bigcup$ S), binding participants to tasks.

Example 2-4. Figure 4 shows a JTAFlow Schema example of student business trip application process. Each box represents an task, the trapezoid one stands for system-activity (This activity does nothing) and each circle represents an state. The lines between entities and tasks mean attributes are readable or modifiable in tasks. According to conditions and route mode, tasks connect states by events. Tasks change attributes, states store and show attributes. Describe about this process has been mentioned in Sect. 2.1.

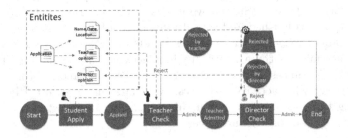

Fig. 4. JTAFlow Schema example: student business trip application process schema

In form of Definition 2-4, we describe Fig. 4 simply as follows:

1. En = {(Application.ApplyInf, userdef, NULL), (Application.TeacherOpinion, bool, false), (Application.DirectorOpinion, bool, false)};
2. T = {(StudentApply, Manual), (TeacherCheck, Manual), (DirectorCheck, Manual), (Rejected, System)};
3. S = {(Start, NULL), (Applied, {ApplyInf}), (RejectedByTeacher, {Apply Inf, TeacherOpinion}), (TeacherAdmitted, {ApplyInf, TeacherOpinion}), (RejectedByDirector, {ApplyInf, DirectorOpinion}), (End, NULL)};

4. C = {(C1, "TeacherOpinion=false"), (C2, "TeacherOpinion=true"), (C3, "DirectorOpinion=false"), (C4, "DirectorOpinion=true")};
5. R = {(R1, Sequence, XOR), (R2, Split, OR), (R3, Jion, XOR)};
6. Ev = {(E1, Start, NULL, R1), (E2, StudentApply, NULL, R1), (E3, Applied, NULL, R1), (E4, TeacherCheck, C1, R2), (E5, TeacherCheck, C2, R2), (E6, TeacherAdmitted, NULL, R1), (E7, RejectedByTeacher, NULL, R1), (E8, DirectorCheck, C3, R2), (E9, DirectorCheck, C4, R2), (E10, RejectedByDirector, NULL, R1), (E11, Rejected, NULL, R1)};
7. P = {(Student, Role), (Teacher, Role), (Director, Role)};
8. M_{EnTS} = {(ApplyInf, StudentApply), (ApplyInf, TeacherCheck), (ApplyInf, DirectorCheck), (TeacherOpinion, TeacherCheck), (DirectorOpinion, DirectorCheck), (ApplyInf, Applied), (TeacherOpinion, RejectedByTeacher), (TeacherOpinion, TeacherAdmitted), (DirectorOpinion, RejectedByDirector)};
9. M_{EvTS} = {(E1, StudentApply), (E2, Applied), (E3, TeacherCheck), (E4, RejectedByTeacher), (E5, TeacherAdmitted), (E6, DirectorCheck), (E7, Rejected), (E8, RejectedByDirector), (E9, End), (E10, Rejected), (E11, End)};
10. M_{PTS} = {(Student, Start), (Student, StudentApply), (Student, Applied), (Teacher, TeacherCheck), (Student, RejectedByTeacher), (Student, TeacherAdmitted), (Director, DirectorCheck), (Student, RejectedByDirector), (Student, End)}.

Definition 2-5. A JTAFlow Instance is a tuple(PI, TI, SI, W, EI, AI), where

1. PI is a set of process instances, each instance is a tuple(PIid, Pid, PIName, Starterid), PIid is an unique process instance ID, Pid is process ID, PIName is name of process instance, Starterid is id of process instance starter;
2. TI is a set of task instances, each instance is a triple(TIid, Tid, PIid, TState), TIid is task instance ID, Tid is task ID, PIid acts as foreign key, TState is state of task such as ready, running, pause, terminate, finished, cancel, working;
3. SI is a set of state instances, each instance is a tuple(SIid, Sid, PIid, Sviewer), SIid is state instance ID, Sid is state ID, PIid acts as foreign key, Sviewer is state viewer;
4. W is a set of workitems, each workitem is a tuple(Wid, TIid, Participant, WState), Wid is workitem ID, TIid acts as foreign key, WState is state of workitem such as ready, running, pause, terminate, finished, cancel, working;
5. EI is a set of tuple (EIid, PIid), which means the mapping between entity instance and process instance.
6. AI is a set of attribute instances, each instance is a tuple(AIid, Aid, EIid, Value), AIid is attribute instance ID, Aid is attribute ID, EIid acts as foreign key, Value is the value of attribute instance.

A JTAFlow Instance example is given in Example 2-6: JTAFlow Snapshot example.

Definition 2-6. A JTAFlow system snapshot is a tuple (Instances, Schemas), where Instances is a set of JTAFlow instances, Schemas is a set of JTAFlow

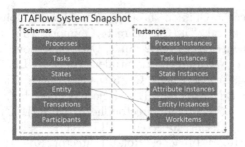

Fig. 5. JTAFlow Snapshot Definition

Table 5. ProcessInstance

PIid	Pid	PIName	Starterid
1	Application	Application1	Bob
2	Application	Application2	Alice

Table 6. TaskInstance

TIid	Tid	PIid	TState
1	StudentApply	1	running
2	StudentApply	2	finished
3	TeacherCheck	2	running

Table 7. StateInstance

SIid	Sid	PIid	Sviewer
1	Applied	1	Bob
2	Applied	2	Alice

Table 8. Workitem

SIid	Sid	PIid	Sviewer
1	Applied	1	Bob
2	Applied	2	Alice

Table 9. EntityInstance

EIid	PIid
1	1
2	2

Table 10. AttributeInstance

AIid	Aid	EIid	Value
1	Application	1	2015/1/30;Bob
2	Application	2	2015/1/30;Alice
3	TeacherOpinion	2	false

schemas. Both of them are defined before. Figure 5 shows JTangFlow snapshot definition, which tells us how instances and schemas work together. After published, schemas are stored in relational tables of engine database, which the engine can recognise. Each instance binds with its schema by recording schema ID.

Example 2-6. Tables 5, 6, 7, 8, 9 and 10 show a JTAFlow Snapshot example based on the JTAFlow schema example in Fig. 4. Table 5 records 2 process instances: Application1 started by Bob and Application2 started by Alice. Task instances are stored in Table 6, which tell us process instance 1(Application1) is in stage of StudentApply, 2(Application2) is in stage of TeacherCheck. Table 7 shows process instance 1 and 2 is in state of Applied. Bob is the viewer of state instance 1, Alice is the viewer of state instance 2. Table 8 shows activity instance 2(StudentApply) is ready for Bob, activity instance 4(StudentApply) is finished and activity instance 5(TeacherCheck) is ready for DoctorA. Mapping between entity instance and process instance is in Table 9. Table 10 records attribute instances. Value of attribute instance TeacherOpinion is initially false as the activity TeacherCheck has not been done. By using PIid, task instances and state instances are connected with process instances. Workitems use TIid to connect with task instances. And attribute instances use PIid because they belong to process instances. In this way, these tables work together and support execution of process instances.

3   Model Mapping Methods

JTangFlow is a executable activity-centric model based on JTangFlow engine, while JTAFlow is not. Mapping from artifact-centric model JTAFlow to activity-centric model JTangFlow is needed to let JTAFlow executable. In this section, we introduce the mapping from JTAFlow to JTangFlow in level of schema. As our aim is to let JTAFlow model executable, the mapping in level of schema meets our need while the mapping in level of instance can be ignored. After analysing differences between JTangFlow and JTAFlow, we summarize following problems:

1. How to transform Entity to Date of JTangFlow? In JTangFlow, data are modeled in a simple form(just put in one line and stored in one table); The engine don't support complex data; furthermore, relation of data or communication about data is ignored in JTangFlow, which is important in real systems;
2. How to transform State to some parts of JTangFlow? JTangFlow doesn't consider the view of data so that State can't be moved to JTangFlow directly.

To address these problems, we propose an algorithm. The main idea of the algorithm is using elements in JTAFlow to describe JTangFlow, adding lacking information and abandoning redundant information such as entity construction, attributes in states.

1. Entity is designed simple in JTAFlow to reduce difficulty of transformation. Data communication between processes or instances is not considered, such as sharing an attribute or attribute instance. Mapping construction of tree(Entity) to vector(Data) is enough so far.
2. Task in JTAFlow is similar to activity in JTangFlow, so we transform the Name and Type of task to activity directly.
3. Compared to task, state do nothing when executed. State looks like gateway that controls route mode from the perspective of execution. It can be regarded as system activity in JTangFlow. The function of data snapshot is abandoned unless support is added to engine. By this way, state in JTAFlow is transformed to system activity in JTangFlow.
4. The same method can be used to transform condition, routemode, event and participant.
5. M_{DA} can be transformed from M_{EnTS} in JTAFlow by replace task or state to activity.
6. So does M_{EA} and M_{PA}.

Algorithm 3-1. The input is a JTAFlow Schema (En, T, S, C, R, Ev, P, $M_{EnTS}, M_{PTS}, M_{EvTS}$). The expected output is a JTangFlow Schema (A, D, C, R, E, P, M_{DA}, M_{EA}, M_{PA}). The mapping method includes the following steps:

1. Initialize JTangFlow Schema, let all elements be empty;
2. For each task(Name, $Type_T$) in JTAFlow.T, add new element activity(Name = task.Name, $Type_A$ = task.$Type_T$) to JTangFlow.A;
3. For each state(Name, Attributes) in JTAFlow.S, add new element activity(Name = state.Name, Type = "System") to JTangFlow.A;
4. For each attribute(path.Name, $Type_A$, InitialValue) in JTAFlow.En, add new element data(Name = attribute.Name, $Type_D$ = attribute.$Type_A$, InitialValue = attribute.InitialValue) to JTangFlow.D;
5. For each $condition_1$(Name, Expression) in JTAFlow.C, add new element $condition_2$(Name = condition1.Name, Expression = $condition_1$.Expression) to JTangFlow.C;
6. For each $routemode_1$(Name, r_1, r_2) in JTAFlow.R, add new element $routemode_2$ (Name = $routemode_1$.Name, r_1 = $routemode_1.r_1$, r_2 = $routemode_2.r_2$) to JTangFlow.R;
7. For each event1(Name, task or state, condition, routemode) in JTAFlow.Ev, add new element event2(Name = event1.Name, activity = event1.task or event1.state, event1.condition, event1.routemode) to JTangFlow.E;
8. For each $participant_1$(Name, $Enum_1$) in JTAFlow.P, add new element $participant_2$ (Name =$participant_1$.Name, $Enum_1$ = $participant_1.Enum_1$) to JTangFlow.P;
9. For each m_{EnTS}(attribute, task or state) in JTAFlow.M_{EnTS}, add new element m_{DA}(data = m_{EnTS}.attribute, activity = m_{EnTS}.task or m_{EnTS}.state) to JTangFlow.M_{DA};
10. For each m_{EvTS}(event, task or state) in JTAFlow.M_{EvTS}, add new element m_{EA}(event = m_{EvTS}.event, activity = m_{EvTS}.task or m_{EvTS}.state) to JTangFlow.M_{EA};
11. For each m_{PTS}(participant, task or state) in JTAFlow.M_{PTS}, add new element m_{PA}(participant = m_{PTS}.participant, activity = m_{PTS}.task or m_{PTS}.state) to JTangFlow.M_{PA};

4 Run JTA in JTang

In this section, we show the implementation of transformation by introducing a JTAFlow example running in JTangFlow engine. The example will be illustrated like this: we design the process by using JTAFlow Designer, get the designed xml, use Algorithm 3-1 to translate it to JTangFlow xml and publish xml into JTangFlow engine.

4.1 Design in JTAFlow Designer

In this subsection, (1) we introduce JTAFlow Designer, (2) use JTAFlow Designer to design a process and (3) get the saved xml.

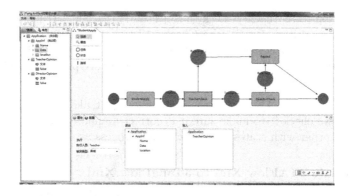

Fig. 6. JTAFlow Designer

JTAFlow Designer is a designer based on JTAFlow model. It contains all elements mentioned in JTAFlow. Users can draw a process in JTAFlow Designer and get the saved JTAFlow xml. JTAFlow Designer has many edit parts: Task, State and Connection. Entity can be built under directory of process. When choosing one element (such as a task), attributes about this task can be set up.

A JTAFlow process can be designed in following steps:

1. Draw taks and states, connect them with connections;
2. Load participants from database, binding participants to tasks and states;
3. Set up route mode of states and tasks;
4. Set up the read and written data in tasks, read and shown data in states;
5. Map data to Entity and build Entity as a tree;
6. Set up conditions involving attributes in Entity;

Here we take student business trip application process as an example:

1. Draw tasks (StudentApply, TeacherCheck, Rejected and DirectorCheck), states (applied, RejectedByTeacher, TeacherAdmitted, RejectedByDirector, Start and End);
2. Binding role student to StudentApply, Applied, RejectedByTeacher, Teacher-Admitted and RejectedByDirector, binding role teacher to TeacherCheck, binding role director to DirectorCheck;
3. Set split mode of task TeacherCheck and DirectorCheck "OR", set jion mode of task Rejected "OR", set other route mode "XOR";
4. Set written data of StudentApply "AppInf", set read data of StudntAp-ply, Applied and DirectorCheck "AppInf", set written data of TeacherCheck "TeacherOpinion", set written data of DirectorCheck "DirectorOpinion", set read data of RejectedByTeacher and TeacherAdmitted "TeacherOpinion", set read data of RejectedByDirector "DirectorOpinion";
5. Build an entity "Application", map "AppInf", "TeacherOpinion" and "Direc-torOpinion" under it;

6. Set condition from TeacherCheck to RejectedByTeacher " TeacherOpinion = false ", set condition from TeacherCheck to TeacherAdmitted " TeacherOpinion = true ", set condition from DirectorOpinion to RejectedByDirector " DirectorOpinion = false ", set condition from DirectorOpinion to End " DirectorOpinion = true ".

After these steps, a JTAFlow process is designed, just like Fig. 6. Before using this process, we check it to make sure it is correct manually(such as tasks must connect each other with states). Then, we save process as a JTAFlow xml file.

4.2   Translate JTAFlow Xml to JTangFlow Xml

In this subsection, we show the transformation of a JTAFlow xml to a JTangFlow xml by our example. In subsection 4.1, we use JTAFlow Designer and design a process of student business trip application. A JTAFlow xml file is produced, part of which content is shown in Fig. 7. And Fig. 8 shows part of the transformed JTangFlow xml.

```xml
<JTAFlow ID="JTAF_32486741584075" xmlns="http://jta.org/flow" xmlns="http://jta.org/flow">
  <Summary>
    <Name>Application</Name>
    <Starter>Student</Starter>
  </Summary>
  <Attributes>
    <Attribute ID="Form_32524482584838">
      <Name>Application</Name>
    </Attribute>
    <Attribute ID="SubForm_32540391095686">
      <Name>AppInf</Name>
    </Attribute>
    <Attribute ID="Entry_32679075939415">
      <Name>Name</Name>
      <Type>string</Type>
    </Attribute>
    <Attribute ID="Entry_32644730389714">
      <Name>TeacherOpinion</Name>
      <Type>bool</Type>
      <DefaultValue>false</DefaultValue>
    </Attribute>
  </Attributes>
  <Entities>
    <Entity AttributeID="Form_32524482584838" Description="application" IsLeaf="false" IsVector="false">
      <children AttributeID="SubForm_32540391095686" IsLeaf="false" IsVector="false">
        <children AttributeID="Entry_32679075939415" IsLeaf="true" IsVector="true"/>
```

Fig. 7. JTAFlow xml example

As we have JTAFlow schema and JTangFlow schema, JTAFlow classes and JTangFlow classes are produced by using XMLBeans. According to **Algorithm 3-1**, a mapping function is written. The transformation is: firstly, we read JTAFlow xml file and set values in class objects; then, We call the function to map JTAFlow class objects to JTangFlow class objects; finally, these JTangFlow class objects are written to JTangFlow xml file. The transformed xml file can be published in JTangFlow engine directly.

4.3   Publish Xml in JTangFlow Engine

In this subsection, we introduce the how to publish the transformed xml file in JTangFlow engine and show the performance of published process. There are

```
<JTangFlow ID="JTangFlow_8Sif6LC5w31417532876037" Name="Application" xmlns="http://jtang.org/flow">
    <Processes>
        <Process ID="Process_05nOW8C8Lt1417532889890" Name="Application" ConRouteType="Bend point">
            <Activities>
                <Activity Type="MANUALACT">
                    <ManualAct>
                        <CommonInfo>
                            <ID>ManualNode_1CAF2vOTx31422360010021</ID>
                            <Name>StudentApply</Name>
                        </CommonInfo>
                        <TaskExecutor>
                            <Executor>
                                <ActExecutor Type="PARTICIPANT">
                                    <Participant>Participant_y58P84qi5J1417532941853</Participant>
                                </ActExecutor>
                            </Executor>
                        </TaskExecutor>
                        <RouteMode>
                            <JoinMode Type="XOR"/>
                            <SplitMode Type="XOR"/>
                        </RouteMode>
                        <StartMode Type="PARTICIPANT">
                        <Datas>
                            <Data ID="Data_o4788550SL1422360010021" Name="_joinValue">
                                <DataType Type="BASIC">
```

Fig. 8. Transformed JTangFlow xml example

two ways to publish the transformed xml file. One is to import it into JTangFlow Designer, another is to invoke JTangFlow API directly. JTangFlow Designer can import xml file only if it comply with JTangFlow schema. What's more, we can modify the imported process as we like. JTangFlow engine opens some API for users to publish process. The transformed xml file can be published while invoking API and passing file to engine.

After published, users can login task manage platform to deal with tasks. A process instance of this example can be as following:

1. Bob logins platform as a student, starts a application process;
2. Bob picks the first task and writes application;
3. DoctorA logins platform as a teacher, reads Bob's application and admits it;
4. DoctorB logins platform as a director, reads Bob's application and rejects it;
5. System task "Rejected" is started, nothing is done and process instance is finished.

5   Related Work

Activity-centric modeling language has developed a lot since 1990s based on petri net [11]. BPEL [1] is an executable business process modeling which widely used in such as JBPM [12], Activiti [4]. The artifact-centric modeling language was initiated by IBM research [7]. Then, more languages such as EZFlow [8], PHILharmonicFlows [9],GSM-based business artifacts [10] are put forward with their advantages in some aspects. Additionally, engines that execute such process models are established [13–15]. As a matter of fact, activity-centric model and artifact-centric model describe business process in different views but the same goal. However, artifact-centric model isn't widely used in industry and has a few executable engines. Translating model from one to another can be a solution. Reference [16] focuses on transformation on control structures but not entity.

Reference [17] provides a set of transformation algorithms to achieve a roundtrip between models. They do it in high level of model which does not aim to support engine running. Additionally, they separate artifact model and object cycle as two models, while we combine them into one model. Reference [18] transforms activity-centric model to information-centric model without attribute consideration. Reference [19] introduces a artifact-centric model called Artiflow and shows that automated translation from ArtiFlow to BPEL is achievable. References [20] focuses on data in business and formulate a data mapping language to map entity to database. Reference [21] proposes a method for translating PNs to GSM models which allows to use existing process mining algorithms for discovering Petri Nets from event logs and generating GSM models from them. They focus on mining and generating while we focus on executing, so they care more about task sequence while we care about data either.

6   Conclusions

In this paper, we introduce an activity-centric model JTangFlow and an artifact-centric model JTAFlow in level of schema, instance and snapshot. Then, we illustrate how to transform JTAFlow model to JTangFlow model by presenting an algorithm. This algorithm tells mapping from JTAFlow model to JTangFlow model in level of schema, especially transformation from Entity and State in JTAFlow. We implement this transformation and execute it by an example in JTangFlow engine. In future work, Entity in JTAFlow will be more strong to support data relationship in or between process instances. We need to develop the algorithm as model changed. Complexity of participant also need to be considered. So we need to guarantee the transformed participant is accepted in JTangFlow engine.

Acknowledgments. This work was supported by National Technology Support Program under grant No.2013BAH10F02.

References

1. Jordan, D., Evdemon, J., Alves, A., Arkin, A., Askary, S., Barreto, C., Bloch, B., Curbera, F., Ford, M., Goland, Y.: Web services business process execution language version 2.0. OASIS standard **11**, 10 (2007)
2. van Der Aalst, W.M., Ter Hofstede, A.H., Kiepuszewski, B., Barros, A.P.: Workflow patterns. Distrib. Parall. Databases **14**(1), 5–51 (2003)
3. Van der Aalst, W.M., Ter Hofstede, A.H.: Yawl: yet another workflow language. Inf. Syst. **30**(4), 245–275 (2005)
4. Rademakers, T.: Activiti in Action: Executable business processes in BPMN 2.0. Manning Publications Co., Shelter Island (2012)
5. Miguel, V., Charoy, F.: Bonita: Workflow cooperative system (2003)
6. Ye, Y., Yin, J., Feng, Z., Cao, B.: Wolf-pack algorithm for business process model syntactic and semantic structure verification in the workflow management environment. In: Services Computing Conference (APSCC), 2010 IEEE Asia-Pacific, pp. 694–699. IEEE (2010)

7. Nigam, A., Caswell, N.S.: Business artifacts: an approach to operational specification. IBM Syst. J. **42**(3), 428–445 (2003)
8. Xu, W., Su, J., Yan, Z., Yang, J., Zhang, L.: An artifact-centric approach to dynamic modification of workflow execution. In: Meersman, R., et al. (eds.) OTM 2011, Part I. LNCS, vol. 7044, pp. 256–273. Springer, Heidelberg (2011)
9. Knzle, V., Reichert, M.: Philharmonicflows: towards a framework for objectaware process management. J. Softw. Maint. Evol. Res. Pract. **23**(4), 205–244 (2011)
10. Hull, R., Damaggio, E., De Masellis, R., Fournier, F., Gupta, M., Heath III, F.T., Hobson, S., Linehan, M., Maradugu, S., Nigam, A.: Business artifacts with guard-stage-milestone lifecycles: managing artifact interactions with conditions and events. In: Proceedings of the 5th ACM International Conference on Distributed Event-Based System, pp. 51–62. ACM (2011)
11. Van der Aalst, W.M.: The application of petri nets to workflow management. J. Circ. Syst. Comput. **8**(01), 21–66 (1998)
12. Cumberlidge, M.: Business Process Management with JBoss jBPM. Packt Publishing Ltd, Birmingham (2007)
13. Heath III, F.T., Boaz, D., Gupta, M., Vaculín, R., Sun, Y., Hull, R., Limonad, L.: Barcelona: a design and runtime environment for declarative artifact-centric BPM. In: Basu, S., Pautasso, C., Zhang, L., Fu, X. (eds.) ICSOC 2013. LNCS, vol. 8274, pp. 705–709. Springer, Heidelberg (2013)
14. Chiao, C.M., Knzle, V., Reichert, M.: A tool for supporting object-aware processes. In: Enterprise Distributed Object Computing Conference Workshops and Demonstrations (EDOCW), 2014 IEEE 18th International, pp. 410–413. IEEE (2014)
15. Yongchareon, S., Ngamakeur, K., Liu, C., Chaisiri, S., Yu, J.: A workflow execution platform for collaborative artifact-centric business processes. In: Meersman, R., et al. (eds.) OTM 2014 Workshops. LNCS, vol. 8842, pp. 639–643. Springer, Heidelberg (2014)
16. Li, D., Wu, Q.M.: Translating artifact-based business process model to BPEL. In: Lin, S., Huang, X. (eds.) CSEE 2011, Part II. CCIS, vol. 215, pp. 482–489. Springer, Heidelberg (2011)
17. Meyer, A., Weske, M.: Activity-centric and artifact-centric process model roundtrip. In: Lohmann, N., Song, M., Wohed, P. (eds.) BPM 2013 Workshops. LNBIP, vol. 171, pp. 167–182. Springer, Heidelberg (2014)
18. Liu, R., Wu, F.Y., Kumaran, S.: Transforming activity-centric business process models into information-centric models for soa solutions. Cross-Disciplinary Models and Applications of Database Management: Advancing Approaches: Advancing Approaches, p. 336 (2011)
19. Liu, G., Liu, X., Qin, H., Su, J., Yan, Z., Zhang, L.: Automated realization of business workflow specification. In: Dan, A., Gittler, F., Toumani, F. (eds.) ICSOC/ServiceWave 2009. LNCS, vol. 6275, pp. 96–108. Springer, Heidelberg (2010)
20. Sun, Y., Su, J., Wu, B., Yang, J.: Modeling data for business processes. In: 2014 IEEE 30th International Conference on Data Engineering (ICDE), pp. 1048–1059. IEEE (2014)
21. Popova, V., Dumas, M.: From Petri nets to guard-stage-milestone models. In: La Rosa, M., Soffer, P. (eds.) BPM Workshops 2012. LNBIP, vol. 132, pp. 340–351. Springer, Heidelberg (2013)

TAR++: A New Process Model Similarity Algorithm Based on the Importance of TARs

Shuhao Wang$^{(\boxtimes)}$, Ming Yin, Zixuan Wang, and Jianmin Wang

School of Software, Tsinghua University,
Beijing 100084, People's Republic of China
{shudiwsh2009,kevinyin1989}@gmail.com,
iamwangzixuan@hotmail.com,jimwang@tsinghua.edu.cn

Abstract. As one of the three elements in enterprises, business process models are their very important asset. An effective process model similarity algorithm is a guarantee for process model retrieval, clustering and classification. In consideration of different deficiencies in existing algorithms, we present a new algorithm called TAR++ which is based on the importance of transition adjacent relations (TARs). The main idea of TAR++ is to describe the relationship of the transitions by adding an importance weight on TARs so as to generate a weighted TAR set and redefine union and intersection operators. In this paper, we introduce the TAR++ algorithm grounded in the Jaccard coefficient and leverages weighted TAR sets. The properties of TAR++ are validated on data sets of SAP, China CNR and DongFang Boiler Group.

Keywords: Unfolding · Behavioral similarity · Transition Adjacency Relation · Importance

1 Introduction

As one of the three important elements of enterprises, the management of process models has always been a big issue. Especially in big group companies, thousands of process models are gathered and form a process repository. How to effectively compare, index and search has become more challenging. The similarity of business models is even significant since a reasonable algorithm can be a guarantee for model retrieval [6]. Many problems remain in existing similarity algorithms, mainly including the dissatisfaction of common similarity properties, high cost of time consumption, and unavailability of specific structures like non-free-choice etc. According to Wang's investigation [10], similarity algorithms should at least satisfy five properties, which will be introduced in the experiment part. The existing algorithms like TAR [11], CF [8], BP [3], PTS [9] cannot satisfy all of the properties. We give a quick introduction about them below.

Zha et al. [11] introduced an algorithm based on Transition Adjacent Relations (TAR for short), which focuses on pair-wise transitions' adjacent relation, namely if transition A and B are sequentially executed in some traces, we call

© Springer International Publishing Switzerland 2015
J. Bae et al. (Eds.): AP-BPM 2015, LNBIP 219, pp. 98–112, 2015.
DOI: 10.1007/978-3-319-19509-4_8

the tuple $\langle A, B \rangle$ is a TAR. Zha used the Jaccard coefficient of TAR sets to represent models' similarity, thus TAR should be a metric. However, TAR cannot handle specific structures like non-free-choice, invisible tasks.

The CF algorithm put forward by van Dongen et al. describes transitions' relations through back-links and forward-links [8]. With these two kinds of links, they compare models' similarity in three levels, namely lexical, syntactic and semantic. But its time consumption proves not ideal, either.

Kunze et al. introduced BP algorithm which defines a weak order to expand the semantics of adjacent relations [3]. The weak order relation includes strict sequential relation, exclusive relation, overlap relation. BP cannot deal with invisible tasks effectively and distinguish behavioral profiles derived from parallel and loop structures.

Professor Wang used principal transition sequences to compute the similarity, namely PTS algorithm [9]. PTS defines three kinds of principal transition sequences. It first calculates the longest common sub-sequences, and then presents the similarity with weighted sum of percentage of sub-sequences. For the concurrent structure with lots of branches, PTS is very time-consuming.

Wang's SSDT makes a matrix with Shortest Succession Distance between Tasks [10]. However, the similarity computation is based on same dimension matrix, which means every time we compute the similarity we need to do the preprocess, which influences the efficiency seriously and makes SSDT not suitable for indexing.

Polyvyanyy et al. proposed another definition of relations between transitions named 4 C spectrum, which includes conflict, co-occurrence, causality and concurrency [5]. However, the hundreds of types of relations in 4 C spectrum is too complex for users to comprehend and hard to compute. Therefore, it is not a practical method in measuring behavioral similarity between models.

In this paper we introduce a similarity algorithm TAR++ based on the importance of transition adjacent relations through adding an important argument or weight to TAR, and thus solve problems that the original TAR algorithm faces. The remainder of the paper is as follows: Sect. 2 introduces the preliminaries including Petri nets, TAR, Complete Finite Prefix etc. Section 3 provides the basic concepts about the importance of TAR, and describes main steps of TAR++ algorithm. After that we prove that TAR++ is a metric and the convergence of it. In Sect. 4 we validate that TAR++ is a metric indeed using data collected from SAP and other enterprises, and compare TAR++ with other algorithms. We conclude the work in Sect. 5.

2   Preliminaries

We give preliminaries in this section, including the basic concepts of Petri nets, the unfolding of Petri nets and semantic presentation techniques, which construct the basis of TAR++ algorithm.

2.1   Petri Net and WF-net

Petri net has been proved to be a powerful modeling language for business process [7]. Workflow net (WF-net for short) is a subset of Petri net. We do our research on WF-net.

Definition 1 (Petri Net). *Let* $P, T \subseteq U$ *be finite and disjoint sets such that* $P \cup T \neq \emptyset$ *and* $P \cap T = \emptyset$, *let* $F \subseteq (P \times T) \cup (T \times P)$. *The tuple* $N = (P, T, F, M_0)$ *is a system, where* P *is the set of places,* T *is the set of transitions,* F *is the set of arcs and* M_0 *as a mapping from* P *to* $\mathbb{N}$ *is the initial marking of the net.* $\mathbb{N}$ *is the set of non-negative integers, i.e.,* $\{0, 1, 2, ...\}$.

Definition 2 (Workflow Net or WF-net). *A Petri net* $N = (P, T, F, M_0)$ *that models the control-flow dimension of a workflow is called a workflow net (WF-net for short), which should satisfy the following conditions:*

- *There is one unique source place* i *such that* $\bullet i = \emptyset$;
- *There is one unique sink place* o *such that* $o \bullet = \emptyset$;
- *Every node* $x \in P \cup T$ *is on a path from* i *to* o.

We skip more details about Petri net and WF-net, but they can be referred in [7]. In this paper, the workflow nets we discuss should be sound, because all the models used in our research have been analyzed as sound and deployed in enterprises for enactments.

2.2   Transition Adjacent Relations

Our metric is based on Transition Adjacency Relations (TAR for short). TAR can be defined in any kind of process formalism and we choose Petri net for demonstration.

Definition 3 (Transition Adjacency Relations). *A Petri net* $N = (P, T, F)$ *is a WF-net. Let* x, y *be two transitions, they are in the adjacent relation* $\rightarrow \subseteq T \times T$, *iff there exists a firing sequence* $\sigma = t_1 t_2 t_3 ... t_n$ *with* $j \in \{1, ..., n-1\}$, *for which* $t_j = x$ *and* $t_{j+1} = y$ *hold. We also say that a tuple* $\langle x, y \rangle$ *is a Transition Adjacent Relation, or TAR. For a given WF-net, all TARs of the model form a set called TAR set.*

Example 1 (An example of TAR). In Fig. 1, the TAR set of the model is $\{\langle T_0, T_1 \rangle, \langle T_0, T_2 \rangle, \langle T_1, T_3 \rangle, \langle T_2, T_3 \rangle, \langle T_3, T_4 \rangle\}$.

2.3   Unfolding of Petri Nets

We use the unfolding techniques [4] to derive basic TARs. Here are the preliminaries for the kind of unfolding used in our algorithm. The detailed description and computation is in J. Esparza's paper [1].

Definition 4 (Complete Finite Prefix). *We say that a branching process* β *of a net system* Σ *is complete if for every reachable marking* M *there exists a configuration* C *in* β *such that:* $Mark(C) = M$ *(i.e.,* M *is represented in* β), *and for every transition* t *enabled by* M *there exists a configuration* $C \cup \{\varepsilon\}$ *such that* $\varepsilon \notin C$ *and* ε *is labeled by* t.

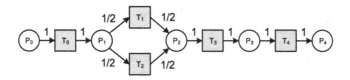

Fig. 1. An example of TAR

3   TAR++ Algorithm

In this section we add importance weights on TARs to derive a new algorithm called TAR++. We first introduce the concepts about the importance of TAR. Then we present how to give each TAR a weight and use the weight to define a new similarity for pair-wise models. Finally we prove this similarity is a metric.

3.1   The Importance of TAR

The TAR algorithm, as mentioned in previous section, takes only pair-wise transition adjacent relations into consideration and thus the importance of TAR does not differ. In consequence, TAR sets cannot reflect the different possibilities of TARs that may fire in concurrent and exclusive structures. For instance, in an exclusive structure with two branches, if the importance of TAR in the trunk is set to 1, then the TAR in each exclusive branch should be half or a user-defined probability since two branches cannot fire them at the same time. We bring in the concepts about the importance of TAR thereby.

Definition 5 (The weight of an Edge). *For a given workflow net, we can add a weight on every edge called the weight of an edge. We also call the weight of an edge adjacent to some certain node the node's weight. The calculation of weights is given in Algorithm 1.*

Example 2 (Weights of Edges). As shown in Fig. 1, we add weights on all edges. The numbers are known as weights of edges. Take place P_1 for example, the weights of P_1 include: the weight of edge $T_0 - P_1$ is 1, while the weight of edge $P_1 - T_1$ is 1/2, and that of $P_1 - T_2$ is 1/2 likewise.

We use three rules to compute the edge weights. Firstly, in a concurrent structure, every transition should occur despite different orders. For every two transitions, their least common ancestor should be a transition if they are concurrent. Therefore, the possibilities of edges connected to a transition should be equal. As for the exclusive structure, only one of the branches can be fired. Here we assume the possibilities of all exclusive branches are equal, but users are allowed to modify these probabilities while keeping the sum of them as one.

Rule 1 (Equality of Weights for Transition Node's Edges). *Given a sound workflow net, the weights of any transition node's edges are equal, including the input and output edges.*

Rule 2 (Equality of Weights for Place Node's Output Edges). *Given a sound workflow net, the weights of any place node's output edges are equal.*

Rule 3 (Conservation of Weights for Place Node's Edges). *Given a sound workflow net, for any place node, excluding the source place or sink place, the weight of its edges should conserve. Specifically, for any node in a WF-net, if its weights of m input edges and n output edges are $\delta_1, \delta_2, ..., \delta_m, \theta_1, \theta_2, ..., \theta_n$ respectively, we can conclude that $\sum_{i=1}^{n} \theta_i = \sum_{j=1}^{m} \delta_j$ holds.*

Example 3 (Application of rules). In Fig. 1, the weights of the two adjacent edges of the transition T_3 are both 1 (Rule 1), while the weights of the two output edges of the place P_1 are both $1/2$ (Rule 2). Besides, the weight of the only input edge of the place P_1 is 1, thus the sum of input edges's weights of P_1 is equal to that of the output edges (Rule 3).

3.2 Computing Weights of a WF-net

If we can compute some node's weights according to the three rules, we call the node's weights decidable. For the whole WF-net, we do the traversal with some specific order. If a node's weights can be decided, we utilize the rules to compute its weights; otherwise, we import new variables on the edges until all the node's weights are marked. After the traversal, we can obtain some equations with unknown variables. Through solving these equations we update the weights. Specifically, the algorithm for computing weights of a WF-net is as follows:

Algorithm 1. Computing the weights of a WF-net

Input WF-net pn

Output weighted WF-net pn

1. Add four nodes and four edges, which include two places Ps and Pn, two transitions Ts and Te and four edges $Ps \rightarrow Ts$, $Ts \rightarrow i$, $o \rightarrow Te$, $Te \rightarrow Pe$ respectively.
2. Initialize a queue $Q = \{Ps\}$; initialize all weights of pn to 0; initialize the weights adjacent to Ps and Pe to 1; initialize all nodes not visited.
3. For each node x in queue Q, circularly visit x till Q is empty. If x has not been visited, calculate the weights of x's edges as many as possible, and then DFS_Search x in pn.
4. When the loop ends, all nodes have been visited and we get a set of linear equations with some unknown weights. Solve the equation and update the weights.

The DFS_Search step is as follows:

Theorem 1. *Through Algorithm 1 we can uniquely and definitely determine all weights of the WF-net. Namely, the weights generated by this algorithm is unique and the algorithm will converge in the end.*

Algorithm 2. DFS_Search

Input WF-net pn, the visiting node x

1. If the weights of x can be calculated with the three rules, do the calculation. Otherwise import n new variables $x_1, x_2, ..., x_n$, and utilize the variables to set the edge's weights.
2. Mark the node visited, and eliminate the node from the queue Q.
3. Let ouput node p be a node adjacent to x.
4. For each output node p, circularly execute:
5. If p is not visited, add p into queue Q, and recursively search p in pn.

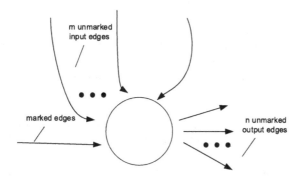

Fig. 2. Example of the marking of a current node

Proof. When we compute the weights of a WF-net, we do the search from the beginning to the end. Meanwhile, we need to import some variables to accomplish the search and finally we solve the linear equations to access the result. According to the three rules, we do linear transformation whether it is a place or transition, so the equations we construct must be linear. Besides, when we import a new variable, it must be like this circumstance shown in Fig. 2: when we come to some place, and the place has m unmarked input edges, n unmarked output edges and some other marked edges whose weights are computed in former search steps. In this circumstance, we need to import m variables in order to represent all adjacent edges' weights, namely, first we add a new variable x to mark all unmarked output edges, and then add another $(m-1)$ variables to mark $(m-1)$ input edges so that the last input edge can be decided using Rule 3. Here the input edges have not been marked yet, so they must be marked through n unmarked output edges' forward traversal, and thus we can construct an equation. Therefore, we get m equations for m variables, and the arguments of these equations should be linearly independent. We solve the equations and a unique solution vector is obtained.   □

3.3 TAR++ Algorithm

For any WF-net PN, if all weights are uniquely decided, we can construct the TAR set with importance coefficients.

Definition 6 (The Importance of TAR). *For any given WF-net PN, we mark an importance coefficient on every TAR in the TAR set. The Coefficient is called the importance of TAR. If all TARs in a TAR set are marked with importance, we say the set is a TAR set with importance, or TAR++ set.*

Algorithm 3. The Computation of the Importance of a TAR

Given a weighted WF-net PN, for any TAR relation $\langle a, b \rangle$, there must exist an intermediate place s connected to transitions a and b simultaneously or a, b are in parallel.

For the former case, they are connected by two directed edges $a \rightarrow s$ and $s \rightarrow b$, whose coefficients are α, β respectively.

For the latter case, there are actually two TAR relations $\langle a, b \rangle$ and $\langle b, a \rangle$. We select one out-edge of a and one in-edge of b for $\langle a, b \rangle$, whose coefficients are α, β respectively. We do similarly to $\langle b, a \rangle$.

We use the minimum number of α and β as the importance value, i.e., $min\{\alpha, \beta\}$.

We construct the TAR set with complete finite prefix technique. Therefore, we do the traversal of the TAR set and compute the importance of TAR according to the weights.

In TAR++ algorithm, TAR++ set can be regarded as a multi-set. The repeatability of the multi-set is the importance. So the operation on the sets should be based on multi-sets.

Definition 7 (Intersection and Union Operation of TAR Sets with Importance). *Given two TAR++ sets named TAR1 and TAR2 respectively, whose elements numbers are m and n. The TARs in these TAR sets are marked tar, and the importance arguments of the TAR++ sets are $\alpha_1, \alpha_2, ..., \alpha_m$ and $\beta_1, \beta_2, ..., \beta_n$ separately. Then we have:*

$$TAR1 \cap TAR2 = \{\delta_i tar_i | \delta_i = min(\alpha_i, \beta_j), \alpha_i tar_i \in TAR1, \beta_j tar_j \in TAR2, tar_i = tar_j\}$$

$$TAR1 \cup TAR2 = \{\delta_i tar_i | \delta_i = max(\alpha_i, \beta_j), \alpha_i tar_i \in TAR1, \beta_j tar_j \in TAR2, tar_i = tar_j\}$$
$$\cup \{\alpha_i tar_i | \alpha_i tar_i \in TAR1 \wedge \forall j, \beta_j tar_i \notin TAR2\}$$
$$\cup \{\beta_j tar_j | \beta_j tar_j \in TAR2 \wedge \forall i, \alpha_i tar_j \notin TAR1\}$$

Therefore, we define the TAR++ similarity based on Jaccard coefficient [2] as follows:

Definition 8 (TAR++ Similarity). *Given two WF-nets N1 and N2, their weighted TAR sets are multi-sets TAR1, TAR2 respectively. Then the TAR++ similarity of N1 and N2 is:*

$$Sim(N1, N2) = \frac{|TAR_1 \cap TAR_2|}{|TAR_1 \cup TAR_2|}$$

Be cautious that TAR1 and TAR2 are multi-sets here, so the intersection and union operation should be based on multi-sets, too.

Example 4 (The computation of TAR++ similarity). Compute the similarity of N1 and N2 in Fig. 3. The TAR++ sets are as follows:

$TAR(N1) = \{1\langle Ts, T_0\rangle, 1\langle T_0, T_1\rangle, 1\langle T_1, T_2\rangle, 1\langle T_2, Te\rangle\}$,

$TAR(N2) = \{1\langle Ts, T_0\rangle, 1/2\langle T_0, T_1\rangle, 1/2\langle T_0, T_3\rangle, 1/2\langle T_1, T_2\rangle, 1/2\langle T_3, T_2\rangle, 1\langle T_2, Te\rangle\}$.

According to the formula of TAR++, the similarity of N1 and N2 is:

$\text{Sim(N1,N2)} = \frac{1+\frac{1}{2}+\frac{1}{2}+1}{1+1+1+1+\frac{1}{2}+\frac{1}{2}} = 0.6$

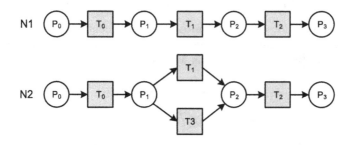

Fig. 3. Calculation of TAR++ similarity

We can also define the TAR++ distance of two WF-nets:

Definition 9 (TAR++ Distance). *Given two WF-nets N1, N2, the weighted TAR++ sets are multi-sets TAR1, TAR2 respectively. We define the distance of the two models as follows:*

$$Distance(N1,N2) = 1 - \frac{|TAR1 \Cap TAR2|}{|TAR1 \Cup TAR2|}$$

Theorem 2. *TAR++ distance is a metric, so that it satisfies reflexivity, non-negative, identity and triangle inequality.*

Proof. Given any two models M1, M2, assume the metric is D. According to TAR++ algorithm, the metric is built on the formula:

$$Distance(M1, M2) = 1 - \frac{|TAR1 \Cap TAR2|}{|TAR1 \Cup TAR2|}$$

Here |TAR1 ⋒ TAR2| and |TAR1 ⋓ TAR2| are operations defined on multi-sets. Consider a mapping $H(MS_1, MS_2)$ which transforms the original multi-sets MS_1 and MS_2 to simple sets TAR_1 and TAR_2, namely:

$$MS_1, MS_2 \xrightarrow{H} TAR_1, TAR_2$$

Assume the arguments of multi-set MS_1 are $\lambda_1, \lambda_2, ..., \lambda_m$, and the arguments of multi-set MS_2 are $\eta_1, \eta_2, ..., \eta_n$, namely:

$$MS_1 = \{\lambda_i tar_{1i}, i = 1, 2, ..., m\}$$

$$MS_2 = \{\eta_j tar_{2j}, j = 1, 2, ..., n\}$$

The specific operation is to calculate a minimum divider E, and then multiply every λ_i and η_j in order to make all coefficients positive integers. The coefficients λ_i and η_j calculated by TAR++ must be rational numbers, and m, n are bounded values, so E must exist and must be unique. We multiply this divider E to every element, so that:

$$MS_1' = \{E\lambda_i tar_{1i} | i = 1, 2, ..., m\} = \{A_i tar_{1i} | i = 1, 2, ..., m\}$$

$$MS_2' = \{E\eta_j tar_{2j} | j = 1, 2, ..., n\} = \{B_j tar_{2j} | j = 1, 2, ..., n\}$$

After this, we divide every tar_{1i} into A_i new elements, namely $tar_{11}^{A1}, tar_{12}^{A2}$, ..., tar_{1m}^{Am}. Then we do the same to every tar_{2j} and divide into B_j new elements, namely $tar_{21}^{B1}, tar_{22}^{B2}, ..., tar_{2n}^{Bn}$. In this way, we have two simple sets:

$$TAR_1 = \{tar_{1i}^{Ai} | i = 1, 2, ..., m\}$$

$$TAR_2 = \{tar_{2j}^{Bj} | j = 1, 2, ..., n\}$$

As for these two simple sets, we can use conclusions and theorems on common simple sets to do the deduction. According to Steinhaus Transform, when given a metric (X, d) and a fixed point $a \in X$, we can define a new metric D':

$$D'(x, y) = \frac{2D(x, y)}{D(x, a) + D(y, a) + D(x, y)}$$

If D is a metric, then the newly constructed D' is a metric likewise. Now assume D is the symmetric difference of the set, then the newly constructed D' is the Jaccard distance. This means the mapped simple set TAR_1 and TAR_2 are in metrics. Since the mapping step only do the marking, the original TAR++ is a metric likewise.   □

3.4   Analysis of Time Consuming

The algorithm of deciding the weights is actually a depth first traversal, and the result is a depth first forest. Since we only do the marking in the process, the time complexity of the step is $O(V + E)$ which is polynomial. After that, we solve equations to obtain the specific number of weights. Suppose when traversal comes to the node with unmarked input edges. Because the traversal is done from the very beginning to the end, these unmarked input edges should be those forward nodes' output edges. We call this kind of unmarked input edges back edges. In the algorithm, if the visiting node is a transition, we should not worry because according to Rule 1, we just go forward and mark the edges. However, if the visiting node is a place, we should import some variables like x. For a model with some node's N back edges, we need to import N variables and thus obtain linear equations with N unknowns. According to Cramer's Rule, the complexity of solving the equations is $O(N!)$. Therefore, the total time complexity is $O(V + E + N!)$. When the back edges are not much, the complexity is nearly polynomial.

4   Experiments and Analysis

4.1   Data Source

Our data are real process models from enterprises like SAP, DongFang Boiler Group and China CNR. These companies are giants of enterprises management, heavy industry etc. The models from the companies have common features of large enterprises, and meanwhile have their own business characteristics. Therefore the data collected is worth researching on.

4.2   Validation of TAR++

As a metric, TAR++ should satisfy its basic property. According to our results, non-negativity and reflexivity are obvious. As for the triangle inequality, there are $C_{10}^3 \times 3 = 360$ groups to be validated in total. We validate that all the 360 groups satisfy triangle inequality. Be limited by space restriction, we won't show the detailed result.

4.3   Comparison of Five Similarity Properties

We mentioned the five properties of similarity algorithms in Sect. 1. We validate TAR++ and other algorithms with these properties in this section.

Property 1 (Sequential structure drift invariance). As shown in Fig. 4, given a sequential WF-net, wherever the new transition is inserted, the similarity of the new model and the old one does not change. Namely, $Sim(N1, N2) = Sim(N1, N3) = Sim(N1, N4)$.

Experiment Models. As shown in Fig. 4, make the number of transitions be 10 in $N1$, then add a new transition to 11 intervals including the beginning and the end. Thus we obtain $N2 - N12$, totally 11 models.

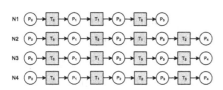

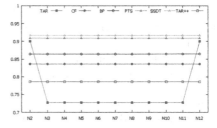

Fig. 4. Sequential structure drift invariance

Fig. 5. Comparison of algorithms with Property 1

The similarity results are shown in Fig. 5. We find that as the position change of the newly added transition, the similarity does not change and thus prove TAR++ satisfies Property 1. Besides, CF, BP, PTS, SSDT also satisfy the property while TAR does not.

Property 2 (Negative correlation by conflict span). As shown in Fig. 6, given a WF-net, as the span of new exclusive branch increases, the similarity of the new model and the old one decreases. Namely, $Sim(N1, N2) > Sim(N1, N3)$.

Experiment Models. As shown in Fig. 6, make the number of transitions be 10 in $N1$, then add a new exclusive branch which step over T_0, then T_0, T_1, and then T_0, T_1, T_2 ... likewise until it step over all 10 transitions. Thus we obtain $N2 - N11$, totally 10 models.

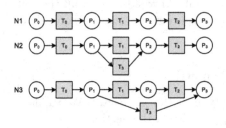

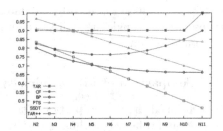

Fig. 6. Negative correlation by conflict span

Fig. 7. Comparison of algorithms with Property 2

The similarity results are shown in Fig. 7. We find that as the span of the newly added exclusive branch increases, TAR++ similarity decreases and thus prove it satisfies Property 2. Besides, PTS, SSDT also satisfy the property while others do not. TAR still has problem at the beginning and the end. CF is not steady with the span increase. BP is a peculiar algorithm, for although it decreases as the span increase, actually this is a specific example. We calculate the BP similarity of the specific 3 models in Fig. 6, and the results are shown in Table 1.

From Table 1 we find that in this dataset the BP similarity of $N1$ and $N3$ is greater than that of $N1$ and $N2$, which proves BP does not satisfy Property 2.

Property 3 (Unrelated task regression). As shown in Fig. 8, given a WF-net, add new branches to the same model in the original model, the more the branches are added, the less the similarity of the new model and the old one is. Namely, $Sim(N1, N2) > Sim(N1, N3)$.

Experiment Models. As shown in Fig. 8, add new branches to T_2 in $N1$. Add one branch as $N2$, two branches as $N3$, and so on. Thus we obtain $N2 - N11$, totally 10 models.

Table 1. Similarity of models in Fig. 6

	$Sim(N_1, N_2)$	$Sim(N_1, N_3)$
TAR	0.500	0.667
CF	0.722	0.744
BP	0.460	0.525
PTS	0.889	0.778
SSDT	0.750	0.688
TAR++	0.600	0.500

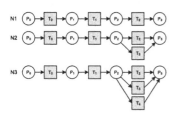

Fig. 8. Unrelated task regression

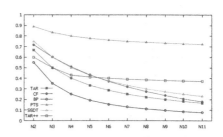

Fig. 9. Comparison of algorithms with Property 3

The similarity results are shown in Fig. 9. We find that as the branches added to T_1 increase, the TAR++ similarity decreases, which proves TAR++ satisfies Property 3. Other algorithms also satisfy the property.

Property 4 (Negative correlation by loop length). As shown in Fig. 10, given a WF-net, add a new loop branch to the original model, the more the span of the branch is, the less the similarity of the new model and the old one is. Namely, $Sim(N1, N2) > Sim(N1, N3)$.

Experiment Models. As shown in Fig. 10, make the number of transitions be 10 in $N1$, then add a new loop branch which step over T_1, then T_1, T_2, and then T_1, T_2, T_3 ... likewise until it step over T_1 to T_9 totally 9 transitions. Thus we obtain $N2 - N10$, totally 9 models.

The similarity results are shown in Fig. 11. We find that only TAR++, SSDT and BP satisfy the property.

Property 5 (Conflict structure drift invariance) As shown in Fig. 12, given a WF-net for deriving an exclusive structure, wherever the new transition is added to, the similarity of the new model and the old one does not change. Namely, $Sim(N1, N2) = Sim(N1, N3) = Sim(N1, N4)$.

Experiment Models. As shown in Fig. 12, make the number of transitions be 10 in $N1$, then add new exclusive branch to 10 transitions. Thus we obtain $N2 - N11$, totally 10 models.

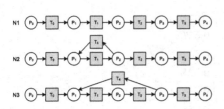

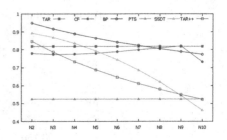

Fig. 10. Negative correlation by loop length

Fig. 11. Comparison of algorithms with Property 4

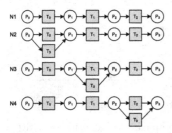

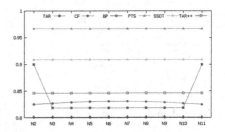

Fig. 12. Conflict structure drift invariance

Fig. 13. Comparison of algorithms with Property 5

The similarity results are shown in Fig. 13. We find that as the position change of the newly added exclusive branch, the similarity does not change and thus prove TAR++ satisfies Property 5. Besides, BP, PTS and SSDT also satisfy the property while others do not.

To sum up, we conclude the algorithms' satisfaction of the properties shown in Table 2. From Table 2, other algorithms cannot satisfy all of the properties except for SSDT and TAR++. However, SSDT is not very flexible as mentioned before, since every time we do the calculation, we need to do the preprocess in order to make the dimensions of two models the same.

4.4   Efficiency Comparison

We have proved that in a model with N back-edges, we import at most N variables so that we obtain all the weights through solving linear equations with N unknowns. The time complexity should be $O(V+E+N!)$. We validate the time complexity with the data sets here. We use 200 models (179 from enterprises and 21 generated) for validation. We calculate the pair-wise similarity of the models and find that CF and PTS is much more time-consuming than others. We cannot reach the results (more than 10000 s) when calculating 200 models using these

Table 2. Validation of algorithms with five properties

Property	Property 1	Property 2	Property 3	Property 4	Property 5
TAR	×	×	✓	×	×
PTS	✓	✓	✓	×	✓
SSDT	✓	✓	✓	✓	✓
CF	✓	×	✓	×	×
BP	✓	×	✓	✓	✓
TAR++	✓	✓	✓	✓	✓

two algorithms. For clarity, we first plot the consuming time of TAR, BP, SSDT, TAR++ in one figure, and then plot CF, PTS, TAR++ in another with log coordinate, as shown in Fig. 14.

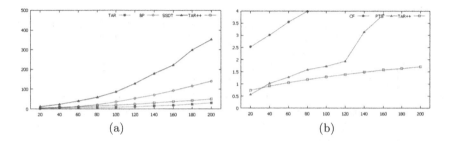

(a)   (b)

Fig. 14. Comparison of different algorithms

From Fig. 14 it is obvious the efficiency of CF and PTS is worse than others. Among them, PTS performs well when the models number is not large, while it jumps to a higher level when the number increases from 100 to 120, because the 20 models added are mostly concurrent structures or loops. PTS does not perform well in these structures. From the comparison of TAR++ with other algorithms, only TAR performs better since we do an extra traversal step in TAR++. Despite of this, TAR++ is more efficient than other algorithms.

5   Conclusion

In this paper we introduce a similarity algorithm called TAR++ based on the importance of transition adjacent relations. As a metric, TAR++ can be used to compare the similarity between pairwise models. The time complexity is $O(E + V + N!)$ while N represents the node which has the most back-edges. As for the evaluation of similarity metric, we put forward five properties including sequential structure drift invariance, negative correlation by conflict span, unrelated task regression, negative correlation by loop length and conflict structure drift invariance. We validate TAR++ and other popular algorithms with these properties and find that TAR++ satisfies all of them. We need to expand the properties of similarity algorithms for future work since there is no standard evaluation framework yet.

Acknowledgement. The research is supported by National Science Foundation of China Project (No. 61472207 & No. 61325008) and the Special Fund for Innovation of Shandong, China (No. 2013CXC30001).

References

1. Esparza, J., Römer, S., Vogler, W.: An improvement of McMillan's unfolding algorithm. In: Margaria, T., Steffen, B. (eds.) TACAS 1996. LNCS, vol. 1055, pp. 87–106. Springer, Heidelberg (1996)
2. Jaccard, P.: Etude comparative de la distribution florale dans une portion des Alpes et du Jura. Impr, Corbaz (1901)
3. Kunze, M., Weidlich, M., Weske, M.: Behavioral similarity – a proper metric. In: Rinderle-Ma, S., Toumani, F., Wolf, K. (eds.) BPM 2011. LNCS, vol. 6896, pp. 166–181. Springer, Heidelberg (2011)
4. McMillan, K.L., Probst, D.K.: A technique of state space search based on unfolding. Formal Methods Syst. Des. **6**(1), 45–65 (1995)
5. Polyvyanyy, A., Weidlich, M., Conforti, R., La Rosa, M., ter Hofstede, A.H.M.: The 4C spectrum of fundamental behavioral relations for concurrent systems. In: Ciardo, G., Kindler, E. (eds.) PETRI NETS 2014. LNCS, vol. 8489, pp. 210–232. Springer, Heidelberg (2014)
6. Shafer, S.M., Smith, H.J., Linder, J.C.: The power of business models. Bus. Horiz. **48**(3), 199–207 (2005)
7. van der Aalst, W.M.P.: The application of petri nets to workflow management. J. Circ. Syst. Comput. **8**(01), 21–66 (1998)
8. van Dongen, B.F., Mendling, J., van der Aalst, W.M.P.: Structural patterns for soundness of business process models. In: 10th IEEE International Enterprise Distributed Object Computing Conference, EDOC 2006, pp. 116–128. IEEE (2006)
9. Wang, J., He, T., Wen, L., Wu, N., ter Hofstede, A.H.M., Su, J.: A behavioral similarity measure between labeled petri nets based on principal transition sequences. In: Meersman, R., Dillon, T.S., Herrero, P. (eds.) OTM 2010. LNCS, vol. 6426, pp. 394–401. Springer, Heidelberg (2010)
10. Wang, S., Wen, L., Wei, D., Wang, J., Yan, Z.: Ssdt-matrix based behavioral similarity algorithm for process models. Comput. Integr. Manuf. Syst. **19**(8), 1822–1831 (2013)
11. Zha, H., Wang, J., Wen, L., Wang, C., Sun, J.: A workflow net similarity measure based on transition adjacency relations. Comput. Ind. **61**(5), 463–471 (2010)

Process Mining

Material Movement Analysis for Warehouse Business Process Improvement with Process Mining: A Case Study

Mahendrawathi ER[✉], Hanim Maria Astuti,
and Ika Rakhma Kusuma Wardhani

Information Systems Department, Faculty of Information Technology,
Institut Teknologi Sepuluh Nopember, Surabaya, Indonesia
mahendra_w@is.its.ac.id

Abstract. Process mining is a technique to model and analyze business process based on traces of activities performed and stored in the database of any information systems being operated by the company i.e. event logs. This paper aims to add literature on the implementation of process mining in warehouse management process. First, extraction of all related activities from LTAK and LTAP tables of SAP Warehouse Management module is conducted. The event log are processed with Heuristic Miner Algorithm in PROM. The output of the mining is the actual business process model. The analysis of the output discover deviation from the standard procedure set by the company with the presence of additional activities and nonstandard flow of materials. The bottleneck analysis shows that the material spent a long time in the high rack and transferred between high racks. The average lead time from material receipt until issued is slightly higher than the target set the company. Further analysis of the issues revealed root-cause of the problems which are insufficient warehouse related capacity such as forklift, racks and staffs to serve incoming materials. Finally, recommendations are provided to deal with the warehouse management issues facing the case companies.

Keywords: Material movement · Warehousing · Process mining · Warehouse management · SAP

1 Introduction

The shift from a functional to a more integrated view in organizations has contributed to increasing attention in business processes. As suggested by [1], managers that take a strictly functional view of the organization face the risk of creating a functional silos. Business process perspective allows the company to focus on a series of activities across different functions in order to achieve their goals and provide value to the customers. Scholars [2, 3] also suggested that organization attempting to improve processes as a basis for gaining efficiency and satisfying their customers showed better overall performance compared to non-process-oriented organizations.

Business process management has received increasing attention recently by different communities notably business administration and computer science communities [4].

© Springer International Publishing Switzerland 2015
J. Bae et al. (Eds.): AP-BPM 2015, LNBIP 219, pp. 115–127, 2015.
DOI: 10.1007/978-3-319-19509-4_9

Business administration communities put more emphasis on the improvement of efficiency and effectiveness of business processes within the companies [5]. Computer science communities focus to enhance the performance of tools or software to support managerial aspirations in improving business processes.

According to [2] BPM can be viewed as continuous cycles comprising several phases of process identification, discovery, analysis, redesign, implementation, monitoring and controlling. Information Technology, particularly Information Systems, plays an important part in business process management since more and more business processes within organizations are supported by information systems [4]. Increasing number of companies in today's business environment implement Enterprise Systems such as ERP, SCM and CRM to conduct their business processes. ERP systems as the first generation of Enterprise Systems are considered to offer great benefits to automate and integrate the business processes of an enterprise [6].

While IT plays a significant role in improving business process, the entire business process management cycle must be carried out continuously in order to achieve maximum value for the company. In other words, the current business process must be evaluated to measure the performance and identify further improvement. Although ERP systems are developed based on best practice, the implementation in real life can deviate from the standard procedure.

Process mining is introduced by Aalst et al. [7] as a technique to model and analyze business process based on traces of activities performed and stored in the database of any information systems being operated by the company i.e. event logs. Process mining enables companies to understand how their business processes are actually conducted. The actual business process itself is presented in a visible and easy to understand form. Based on the actual condition of the current business process company can identify discrepancy from the standard or ideal business process (if any) and identify potential cause and solutions to the discrepancy.

Process mining has received a lot of attention in the literature. Research on developing algorithms to model the event logs has been done by [8–10]. Previous works done by [11, 12] proposes several measure to evaluate the performance of process mining algorithms. The works of [13, 14] focus on identifying challenges in mining supply chain process and the solution identification using RFID.

Several authors also presented the implementation of process mining to solve real case. Jans et al. [15] presented their work with respect to internal fraud mitigation in a SAP based Procurement. Rebuge and Ferreira [16] implement process mining and several analysis in healthcare environment. De Weerdt et al. [17] presented a methodology to implement process mining in financial organization. Our previous works implement process mining to model real case business processes including incoming materials and production planning [18, 19]. Nevertheless, each type of industry have different processes. Therefore, more studies need to be done to provide a comprehensive view of the success of process mining in solving real cases.

The work presented in this paper uses process mining technique to model and analyze warehouse management business process in a manufacturing company in Indonesia. Warehousing plays a significant role in any manufacturing companies as it deals with various business processes related to the movement and storage of materials within the company along the supply chain [20]. A good warehousing management is believed to result in a sustainable supply chain management. However, there are several challenges occurred in warehousing process including uncontrolled material movements due to the issue of information transfer [21]. The case company uses Warehouse Management (WM) module from SAP to manage the entire warehouse starting from receive, inspection, handling, storing and transferring raw materials. The case company admits several issues in the management of raw materials in the warehouse. However, there have not been any evaluation on the current business processes.

The study reported in this paper extracts event log from WM module of SAP and applies process mining algorithm to obtain the process model. Following the modelling, a two-stage analysis are conducted. First, results from the model are used to: (1) investigate conformance to the standard operating procedure enforced in the case companies (2) obtain key performance in terms of average holding time. Findings from the first stage is followed by a more in-depth analysis with the case company to obtain root-cause of the problems.

The contribution of this paper is twofold. First, this paper provides insights on the implementation of process mining especially in the warehouse business process in a real case. This can add more range in the literature on the applicability of process mining for business process improvement in the context of a real case. Secondly, it also provides practical implication for the company by highlighting issues in the warehouse management process and proposing recommendations to address the problems.

This paper is organized as follow. First, overview on the literature on process mining are discussed. This is followed with a brief description on warehouse management business process in the case company. Section 4 describes the methodology of the study. Results and analysis obtained from the process mining implementation are described in Sect. 5. Section 6 discusses root-cause of the issues identified in Sect. 5 as well as recommendations to deal with the issues. Finally, a brief concluding remarks are given in the penultimate section.

2 Literature Study

Companies typically execute dozens of business processes in their day-to-day operation. Since a business process is a collection of activities conducted by a company to achieve an expected outcome [22], a good information system is ideally designed to support the activities of a company. [23] emphasis the importance of designing a sound business process in order to achieve efficiency in the daily operation. Several techniques to design a business process are introduced to improve the analyst's understanding on how activities in a company are conducted and how to easily document these activities. The activities are usually represented in visualized forms such as flowchart, activity diagram, sequence diagram, Business Process Modeling Notation (BPMN), Petri-net and others.

Following the presence of various techniques to represent a business process, the evaluation of current business processes running in a company has attracted the attention of many scholars. Process mining technique [7] emerges to serve such purpose. Process mining is a technique used to explore the business processes of a company based on the event logs of information systems being used in the company.

This technique enables managers to evaluate their existing business processes. A group of researchers [7] has developed a tool for mining business processes called PROM [24].

An information system typically has a history or log that records all activities conducted by users in a certain time. Process mining makes use of the event logs of information systems as an input to visualize the business processes in the form of Petri-net. The event logs must be converted into certain format, usually in MXML [25] or XES [26] to be processed using PROM tool. Several tools developed by Fluxicon [27] such as Nitro and Disco can be used for the conversion purpose. ·

One of the main challenges in implementing process mining in a real case, is that event logs cannot be retrieved instantly from an information systems. Not every company activate their event logs as it requires large disc space and potentially slow down the systems. In this case, event logs must be extracted from the systems' databases, which stored all activities performed by users. A tool called XESame [28] is developed in 2010 to facilitate the extraction process from any database into logs. These logs are then converted into the required format to be modeled and analyzed using PROM tool.

Several process mining algorithms are developed to model the business processes in a visualized form. The algorithm aimed to construct causalities from various sequential activities. PROM tool has provided several algorithms with different characteristics. Alpha and Alpha++ algorithms are one of the earliest process mining algorithms. These algorithms are not sensitive to noise and missing values [7, 10]. In addition, frequency is not considered in these algorithms. Even though Alpha++ is better than Alpha algorithm in term of its ability to detect implicit dependency [10], the performance especially in terms of fitness are not as good as other algorithms. Genetic Algorithm (GA) is another algorithm available in PROM tool. This algorithm is quite sensitive to noise and it appears to be unstable in several circumstances especially when there are several branches in a model [8]. Heuristic Miner algorithm can better handle noises and missing values. This algorithm is also better than GA since it counts the frequency of relation between activities in order to construct the dependency value between the activities [9]. Heuristic miner is quite suitable to model the actual business processes in order to assess conformance with the standard or the pre-defined business processes.

3 Warehouse Management in the Case Study

The case company used in this research is a subsidiary of an international shoe manufacturer located in East Java, Indonesia. The company produces different types of leather shoes for women, men, kids and sports for two selling season autumn-winter and spring-summer. The company procures materials such as rawhides, reinforcement, yarns, shanks, etc. from suppliers in Europe, China, Indonesia, and several other countries. The company implements SAP ERP to conduct its numerous business processes including warehouse management. All the processes in the raw materials warehouse are managed using the Warehouse Management (WM) Module in SAP starting from receiving goods (good receipt), storing goods (good storage) and issuing goods (good issue). Each incoming material is checked in terms of quantity and quality before being put in a storage. The materials are issued according to production order requested by Material Requirement Planning (MRP). The standard procedure for materials movement from good receipt until good issue in the case company warehouse is depicted in the following figure.

According to Fig. 1, the standard procedure of materials movement begins with good receipt in the warehouse. The materials are typically ordered in big batch but they can arrived and received at the warehouse in several batches. Every materials received can then be divided into smaller batches for Quality Inspection (QI). A transfer order document is generated for each material receipt containing information on material code, size, quantity and date received. Batch of materials that have passed QI are moved into the storage and organized based on a certain category. There are two types of storage: high rack and picking rack. High rack is used to store materials that passed the QI or materials that must wait for production. Picking rack is a storage used to store materials with more immediate production schedule. Batch of materials that fail to pass QI are stored in a storage named blocked stock. Materials are then issued for production process according to production order from Material Resources Planning.

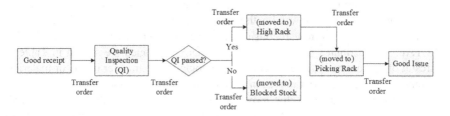

Fig. 1. The standard procedure for materials movement in the case company warehouse

4   Methodology

Several stages are conducted to analyze the material movement in warehousing using process mining technique as shown in Fig. 2.

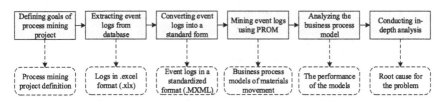

Fig. 2. The methodology to analyze the actual business processes of material movement. The rectangle indicates a process while the rounded rectangle with dashed line indicates the output of a process.

4.1 Process Mining Goals, Scope and Focus

The case company has never analyzed the existing warehouse business process and its performance. However, they already have a standard procedure for material movement as shown in Fig. 1. One of the key performance indicators for a good warehouse is average holding time of materials. The company set a target of 30 days for material holding time. The main goal of the process mining implementation is therefore to identify conformance to standard procedure of material movement in the case company warehouse and measure the performance of the warehouse management. The scope of the process mining implementation is the material movement starting from good receipt until it is issued from the warehouse. A control flow perspective is taken. The main focus of the analysis is on the elapsed time from good receipt into until good issued from the warehouse.

4.2 Database Extraction and Event Logs Conversion

The process mining definition described above is used to guide the database extraction and event log conversion. The log must be extracted from the SAP tables because the event log feature in SAP is not activated. As stated above, a control flow perspective is used in this case and thus the event log must contains case, activities, and timestamp.

According to the interview with the person in charge with the SAP in the case company, all transactions in warehousing processes are recorded in two tables: LTAK and LTAP. The attributes from the two tables that is extracted are shown in Table 1. The result of this extraction is an excel file that contains the transfer order for materials movement. Since the material moves in different batch size, the case in the event log is the smallest batch of material that moves from good receipt, quality inspection, high rack, pick rack until good issue. In order to enable tracing the smallest batch, the Case ID is generated from the combination of material ID, material size, source storage unit and the order of material movement.

Table 1. The attributes in the SAP tables.

No	Attribute table
1	BDATU (Creation Date of Transfer Order)
2	WDATU (Date of Good Receipt)
3	VLPLA (Source Storage Bin)
4	NLPLA (Destination Storage Bin)
5	VSOLM (Source target Quantity)
6	Grid Value
7	Material Description
8	BESTQ (Stock Category in the WMS)
9	Material ID
10	VLENR (Source storage unit)

The next step is to obtain the activities of the event log. Based on the standard procedure there are six activities in the warehouse activities to be captured including: (1) *Good Receipt*, (2) *Quality Inspection*, (3) *Block Stock*, (4) *Good Issue*, (5) *Create TO High Rack*, (6) *Create TO Picking Rack*. The first four activities are identified from the movement type attribute. But, the last two activities i.e. *Create TO High Rack* and *Create TO Picking Rack* are identified from attribute source storage bin. If the source storage bin is pick rack then the activity is *Create TO Picking Rack*. As the data is perused, two other additional source bins are identified. When this is confirmed to the company staff, they said that in reality there are two additional activities in warehouse process i.e. *Create TO Lantai* that stands for creation of transfer order to the floor and *Create TO Antar High Rack* that stands for creation of transfer order between high racks. The occurrence of these additional activities will be discussed further in the analysis section. Finally, the time stamp is obtained from creation date attribute.

The excel file is then converted into a standardized format of event logs using Nitro software. According to the conversion, there are 15 identified scenarios of material movement that can be modeled using process mining technique. The next stage is the conversion from excel format into MXML.

4.3 The Mining of Event Logs

Having a standardized format in MXML, the event logs are then processed using process mining tool PROM. The algorithm used for modelling is Heuristic Miner since the aim of this study is to obtain an ideal business process model of material movement, i.e. a model that most closely meets the standard procedure set by the company. This means that the model will be constructed by neglecting activities with small frequency. The process model of material movement is described in the following figure.

The model is analyzed according to the performance measurement provided by the PROM tool. The fitness, precision and structure are 0.9993, 0.9365, and 1 consecutively. This shows that the algorithm can capture the characteristics of the event log into the business process model. The results from the model can then be used for further analysis.

5 Result and Analysis

Based on the process mining results, three analysis are conducted. The first is analysis of conformance of the actual business process models and the standard procedure. The second analysis is to identify the length of time the materials are held in stock. The final analysis is dotted chart analysis to identify the company rule in material movement.

5.1 The Conformance of the Actual and the Standard Procedure

Comparison between the process models depicted in the Fig. 3 and the standard procedure described in Fig. 1, shows that the material movement have not entirely conform to the standard procedure. Three deviations in the warehousing business processes are identified:

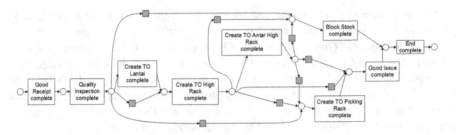

Fig. 3. The Petri Net model of material movement in the case company.

1. **Additional Activities.** The model highlights the presence of additional activities that have not been defined in the standard procedure i.e. *Create TO Lantai* and *Create TO Antar High Rack*, with the number of frequency of scenarios containing both activities is nearly 53 %. This means that more than half of the case analyzed do not conform to the standard procedure.
2. **Nonstandard Flow of Activities.** There is a high possibility to issue goods directly from other activities rather than *Create TO Picking Rack*. Ideally, a material is issued after it has been stored in a picking rack. However, the analysis shows that around 50.14 % of the materials issued without being preceded by *Create TO Picking Rack*.
3. **Quality Issues.** There is also a possibility for Blocked stock to occur after materials stored in high rack. Ideally, all materials stored in high rack have passed the QI process while those that fail the QI are stored in blocked stock. However, the result shows that there are several materials that have passed the QI and stored in the high rack but is later moved into blocked stock.

5.2   Length of Materials in the Stock

In order to have an overview of the length of time materials stored in the warehouse a bottleneck analysis is done in PROM. The result can be seen in Fig. 4 which show that the bottleneck activities (the longest waiting time relative to other activities) occurred on the movement of materials from cr*eate TO High Rack* and from *Create TO Antar High Rack.*

A more detailed depiction on the longest waiting time from *Create TO Antar High Rack* and from *Create TO High Rack* can be seen in Table 2.

Activities *Create TO Antar High Rack* on average took 14 days, while *Create TO High Rack* on average took 15 days. In total, the average time from materials receipt until they are issued are 31 days.

5.3   Dotted Chart for FIFO Analysis

The final analysis uses Dotted Chart analysis in PROM to review the rule for material movement. In the case company, the same materials can arrived in separate batches.

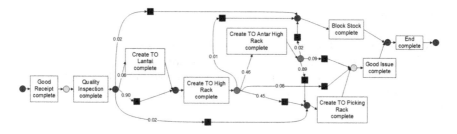

Fig. 4. The bottleneck activities in the warehouse.

Table 2. The length of materials movement in warehousing and the waiting time from two activities i.e. *Create TO* Antar *High Rack* and *Create TO High Rack* to the next process. Max refers to maximum time, Min refers to minimum time and Avg refers to average time.

Waiting time from *Create TO Antar High Rack* to the next process			Waiting time from *Create TO High Rack* to the next process			Time required from *Good Receipt* until *Good Issue*		
Max	Min	Avg	Max	Min	Avg	Max	Min	Avg
217 days	0.02 days	14 days	128 days	1 h	15 days	220 days	0.29 days	31 days

Ideally, in order to meet the 30 days average holding time target, batch of material that arrive first must be taken out first i.e. FIFO rule. Dotted chart shows each instance in a row and dots that represents activities from the event log in different colors.

An example of Dotted Chart with 77 instance of material SHANK PLASTIC TOUCH 50 is shown in Fig. 5a, while an ideal condition if the movement follow FIFO policy is shown in Fig. 5b. The figures show activities from *Good Receipt – Quality Inspection – Create TO High Rack – Create TO Picking Rack – Good Issue* as dots in different colors. FIFO rule can be identified by perusing the first and last activities. It is clear that the material is not taken out on a FIFO basis as some of the instance that is received earlier (blue dot) is taken out later (red dot). Further analysis of another 9 materials all show that they are not following the FIFO Policy.

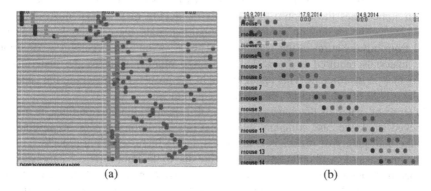

(a)   (b)

Fig. 5. Dotted chart of material SHANK PLASTIC TOUCH 50 (a) and the Ideal Dotted Chart if the material follow FIFO policy (b)

6 Discussion

Results from the process mining provided in the previous section reveal five issues in warehouse management process of the company. In order to understand the root-cause of the issues an interview are held with representatives from the company which include warehouse manager and staffs. The interview results are summarized in Table 3.

Table 3. Root-cause of the warehouse issues.

Issues	Root-cause
Additional activities	Limited capacity of forklift, staff and storage
Nonstandard flow of activities	Materials in picking rack not available in the right quantity
Quality issues	Some materials quality can only be determined in assembly process
Bottleneck in some activities	Capacity of picking rack, production plan changes, early delivery of materials
Materials not taken out in FIFO	No emphasis on such rule

As shown in Table 3, it was found that the deviation from standard mostly due to capacity issues. First, additional activities of *Create TO Lantai* occurs as a result of: (a) limited number of forklift to move materials from QI to High Rack, (b) the limited number of staffs working in warehousing unit and (c) for certain period—usually in the transition from Summer Spring to Autumn Winter, the storage is normally full of capacity. To overcome this problem, the materials are temporarily placed on the floor. In addition to the aforementioned causes, *Create TO Antar High Rack* usually occurs because the storage is already full of materials so that the new incoming materials are temporarily stored in a wherever there is available space in high rack before moved to the original designated space in the high rack.

Another issue is the fact that the issuing of material does not always follow the standard. The procedure states that material must be issued from picking rack. However, in some cases the amount of materials in picking rack is inadequate for production. Since the production requires additional materials, these materials are usually issued from any location in the warehouse. The deviation also occurs in a condition where materials required to be issued in a large number whereas the materials in picking rack is insufficient. This again highlight the problem of matching the capacity of storage with the demand i.e. production requirements.

In terms of materials that have passed the QI inspection but later on must be moved to blocked stock the case company reveal that this condition happens because sometimes the bad quality of materials is identified during the assembling process.

The presence of two bottleneck activities are due to the imbalance capacity between high rack and picking rack. Picking rack is used to store materials which are ready for production. The picking rack has small capacity in comparison to high rack. Naturally, this means that some materials are stored in high rack longer relative to other activities. However, the bottleneck may also occur due to the frequent changes in production planning. A change in production plan influence the operation of the warehouse. Materials can be canceled if there is a change in the production plan. As a result, the materials are stored longer in the warehouse. Finally, materials may stay longer in high rack because it is delivered earlier than expected production schedule. This means the materials must be stored in the warehouse until the production is started.

Dotted chart analysis shows that the warehouse do not follow the FIFO policy in the raw materials. As a result the average holding time for the materials are all above the standard company of 30 days. When it is confirmed to the person in charge in the warehouse they argued that they have not emphasis on FIFO rule for material movement. This is because the difficulty to trace the materials that arrives first and also due to the aforementioned issues where the materials cannot always be put in the designated place.

Overall, the results from the study highlight the challenges faced by the case company in managing its warehouse business process. Several recommendations can be made to deal with the issues. Clearly, capacity in terms of forklift, high rack, pick rack, and staffs are the main issues that need to be resolved. The company must try to better match the resource capacity with the demand i.e. incoming materials as well as production requirement. The real challenges appears that the production plan frequently changed. This most likely creates the problem of putting the materials on the floor, transferring between high racks and issuing materials not from the pick rack. When the demand is manageable then the warehouse staffs can ensure a better placement of materials and treat them in a FIFO basis and eventually enable them to meet the target average material holding time of 30 days.

7 Conclusion

To conclude, this paper presents the implementation of process mining to model and analyze the materials movement in raw material warehouse. Since the event logs of the company case study are not activated, an extraction of all related activities from LTAK and LTAP of warehouse management module is conducted. The event log are processed with Heuristic Miner Algorithm in PROM. The output of the mining is the actual business process model. The analysis of the output shows that there are three deviations between the actual and the defined business processes, i.e. the presence of additional activities which are not in the standard process, the possibility to skip moving materials to pick rack and the possibility to move material that has been put in high rack to block stock. The analysis also shows bottleneck activities and that the materials are not issued in FIFO basis. As a result the average length between materials received until issued is 31 days, which is slightly above the standard 30 days set by the company. Further analysis of the issues revealed that the main problem in warehouse management is matching the warehouse related capacity (forklift, high rack, pick rack, and staffs). To deal with the issue the company must first attempt to ensure that production plan is stable and do not change too frequently. Once the demand is manageable the warehouse manager can then calculate the real capacity required to satisfy the demand. Finally, this study highlight that process mining can help company to reveal issues in the existing business processes and structure discussion to identify root cause of the problem and find ways to deal with them.

References

1. Wisner, J., Stanley, L.: Process Management: Creating Value Along the Supply Chain. Cengage Learning, Boston (2007)
2. Dumas, M., et al.: Fundamentals of Business Process Management. Springer, Heidelberg (2013)
3. McCormack, K.P.: The Development of a Measure of Business Process Orientation. Paper presented at the European Institute for Advanced Studies in Management: Workshop on Organizational Design, Brussels, Belgium (1999)
4. Weske, M.: Business Process Management: Concepts, Languages, Architectures. Springer, Heidelberg (2007)
5. Cuthbertson, R., Cetinkaya, B., Ewer, G., Klaas-Wissing, T., Piotrowicz, W., Tyssen, C.: Sustainable Supply Chain Management: Practical Ideas for Moving Towards Best Practice. Springer, Heidelberg (2011)
6. Langenwalter, G.A.: Enterprise Resources Planning and Beyond: Integrating Your Entire Organization. St. Lucie Press, Florida (2000)
7. van der Aalst, W.M.P., Weijters, T., Maruster, L.: Workflow mining: discovering process models from event logs. IEEE Trans. Knowl. Data Eng. 16(9), 1128–1142 (2007)
8. de Medeiros, A.K.A., Weijters, A.J.M.M., van der Aalst, W.M.P.: Genetic process mining: an experimental evaluation. Data Min. Knowl. Disc. 14(2), 245–304 (2007)
9. Weijters, A.J.M.M., Medeiros, A.K.A.D.: Process Mining with the Heuristics Miner-algorithm. Cirp Anal. - Manufact. Technol. (2006)
10. Wen, L., van der Aalst, W.M.P., Wang, J., Sun, J.: Mining process models with Non-free-choice constructs. Data Min. Knowl. Disc. 15(2), 145–180 (2007)
11. Rozinat, A., Alves de Medeiros, A.K., Gunther, C.W., Weijters, A.J.M.M., van der Aalst, W.M.P.: Towards an Evaluation Framework for Process Mining Algorithms. Eindhoven University of Technology, Eindhoven (2007)
12. Rozinat, A., van der AAalst, W.M.P.: Conformance Checking of Processes based on Monitoring Real Behavior. Inf. Syst. 33(1), 64–95 (2008)
13. Gerke, K., Claus, A., Mendling, J.: Process mining of RFID-Based supply chains. In: Proceedings of the 2009 IEEE Conference on Commerce and Enterprise Computing, pp. 285-292. IEEE Computer Society (2009)
14. Gerke, K., Mendling, J., Tarmyshov, K.: Case construction for mining supply chain processes. In: Abramowicz, W. (ed.) Business Information Systems. LNBIP, vol. 21, pp. 181–192. Springer, Heidelberg (2009)
15. Jans, M., van der Werfb, J.M., Lybaerta, N., Vanhoofa, K.: A business process mining application for internal transaction fraud mitigation. Expert Syst. Appl. 38(10), 13351–13359 (2011)
16. Rebuge, Á., Ferreira, D.R.: Business process analysis in healthcare environments: a methodology based on process mining. Inf. Syst. 37(2), 99–116 (2012)
17. De Weerdt, J., Schupp, A., Vanderloock, A., Baesens, B.: Process mining for the multi-faceted analysis of business processes—a case study in a financial services organization. Comput. Ind. 64(1), 57–67 (2013)

18. Mahendrawathi, ER., Astuti, H.M.: A case study on process mining implementation in modeling supply chain business process: a lesson learnt. In: 6th International Conference on Operations and Supply Chain Management (2014)

19. Mahendrawathi, ER., Astuti, H.M., Pramitasari, D.: Modeling and analysis of incoming raw materials business process: a process mining approach. Int. J. Comput. Commun. Eng. **4**, (3), 196–203 (2015)

20. Ackerman, K.B.: Practical Handbook of Warehousing, 4th edn. Kluwer Academic Publisher, Norwell (1997)

21. Voortman, C.: Global Logistics Management. Juta and Co Ltd., Cape Town (2004)

22. Davenport, T.H.: Process Innovation: Reengineering Work through Information Technology, p. 337. Harvard Business School Press, Boston (1993)

23. Li, J., Liu, D., Yang, B.: Process mining: extending α-Algorithm to mine duplicate tasks in process logs. In: Chang, K.C.-C., Wang, W., Chen, L., Ellis, C.A., Hsu, C.-H., Tsoi, A.C., Wang, H. (eds.) APWeb/WAIM 2007. LNCS, vol. 4537, pp. 396–407. Springer, Heidelberg (2007)

24. Group, P.M. ProM (2009). http://www.processmining.org/prom/start. Accessed 19 December 2014 Cited 2015

25. Group, P.M. MXML (Mining eXtensible Markup Language) (2009). http://www. processmining.org/logs/mxml. Accessed 15 August 2011 Cited 2015

26. Group, P.M. XES (eXtensible Event Stream) (2009). http://www.processmining.org/logs/ xes. Accessed 16 April 2011 Cited 2015

27. Fluxicon. Process Mining for Professionals (2012). http://fluxicon.com/. Cited 2015

28. Buijs, J.: Mapping Data Source to XES in a Generic Way, in Mathematics and Computer Science. Eindhoven University of Technology, Eindhoven (2010)

An Analytic Framework of Design for Semiconductor Manufacturing

Chia-Yu Hsu[✉]

Department of Information Management and Innovation Center for Big Data
and Digital Convergence, Yuan Ze University, Chungli 32003, Taiwan
cyhsu@saturn.yzu.edu.tw

Abstract. The manufacturing intelligence behind data, process, and system are used to assist extracted from a large amount of manufacturing-related data from many sources, which presents the relevant knowledge to enhance decision quality with operational efficiency. Based on the data-driven view, this study aims to develop an analytic framework of design for semiconductor manufacturing. To demonstrate the proposed framework, a case study in semiconductor manufacturing regarding the layout design of chip size was conducted. The proposed framework can be used to extract useful information and assist effectively process operation by data mining methods with hybrid algorithms to enhance decision quality and production effectiveness.

Keywords: Business analytic process · Yield enhancement · Data mining · Manufacturing intelligence · Semiconductor manufacturing

1 Introduction

Design for manufacturability (DFM), a concept of designing product with consideration among manufacturing, test and assembly simultaneously for ensuring that a product can be manufactured repeatedly, reliably, and cost effectively [1]. DFM is to conceive and refine design alternatives that make the best use of manufacturing capabilities, in terms of the various material processes, tool, equipment, and facilities available realize the design [2].

As the technology is continuously advanced and required functionality is increasing, the time left to engineers for yield enhancement is drastically reduced. In the past, most of the process variations in manufacturing could be solved by improving manufacturing technology. However, in nanometer technology, yield loss is mainly caused by variation of process and device that is increasing hard to be reduced by manufacturing process improvement, yet can be improved by modified IC design. Thus, the designer should consider the potential difficulty and production effectiveness in manufacturing process as designing an IC. However, the IC designers are not experts regarding manufacturing technology, and thus they need to collaborate with manufacturers to obtain the knowledge of the manufacturing process.

DFM scheme should be identified the causal relationship between the designable variables for IC design and manufacturability for wafer fabrication in advance, but some of them are difficult to be formulated directly by physical and mathematical

© Springer International Publishing Switzerland 2015
J. Bae et al. (Eds.): AP-BPM 2015, LNBIP 219, pp. 128–137, 2015.
DOI: 10.1007/978-3-319-19509-4_10

equation. However, huge amounts of data have been recorded during design and manufacturing process, yet meaningful information is latent behind the huge data. The extraction of information can bridge the gap between design and manufacturing to enhance manufacturability. Although a number of studies have addressed modifying physical layout design to improve the manufacturability, little research has been done in mining potentially causal model or rules for constructing the relationships between IC design variables and manufacturing response variable from large databases to assist chip layout design.

This study aims to address an analytic framework of design for semiconductor manufacturing. Through analyzing amount of data for layout design and manufacturing process, data mining can build an empirical model for IC designers to understand the relationship between the designable and response variables while the relationship is difficult to express via a physical formula.

2 Literature Review

2.1 Semiconductor Phase and Data

Design, manufacturing and test are three mainly phases in semiconductor industry as shown in Fig. 1. The IC development processes for designing the integrated circuit (IC) layout includes system design, structural and logic design, transistor-level design and layout design, which are cycle of top-down design and bottom-up redesign [3, 4]. In particular, to ensure the design IC function and performance, the designer must plan the layout with the specific design rules which describe geometric constraints such as the minimum line width, minimum space and interlayer registration [5]. The design rules mainly make that the geometrical reproduction of pattern which can be rightly printed by the mask-making and lithography process, and the interaction between different layers can be solved [6]. Next, after completing of IC layout design, it is taped out to make design pattern on the mask for further manufacturing process.

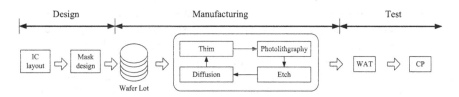

Fig. 1. Semiconductor phase and data type

The wafer fabrication process for producing IC also consists of a lengthy sequence of complex physic-chemical processes involving cycling processes of thin-film deposition, oxidation, photolithography, thin-film etching, and ion implantation on the surface of single crystal silicon wafers layer by layer. During fabrication manufacturing, massive amounts of process data including lot history and tool history are automatically collected. Then, electrical test, i.e., wafer acceptance test (WAT), including the threshold voltage, channel length, channel width, and contact resistance is performed on test keys distributed across the wafer to monitor the characteristics of fabricated ICs at the end of fabrication processes. Then, functionality test, i.e., circuit probing (CP) test, of on each die on the wafer is conducted to distinguish good ones from defected in wafer sort. Finally, the wafers pass all the tests will be sent to assembly to dice up for packaging the good dies and then conduct final test of packaged IC.

Owing to advances of information technologies and new applications, large amounts of data at various stages of design, manufacturing and testing are recorded that can provide a rich resource for data mining and knowledge discovery from database. This data could be relative to IC layout designs, products, processes, equipments, materials, inspections, inventories, sales, marketing and performance data and could include patterns, trends, associations, and dependencies [7]. Data mining is a multi-disciplinary approach for discovering potentially useful patterns from large amounts of data stored either in databases, data warehouses, or other information repositories [8]. For IC design and mask design process, there are two main data will be recorded as follows [6]:

(i) IC layout data: layout data shows the design pattern of IC including chip length, chip width, and the density of layout pattern.
(ii) Design rules: design rules is to define the line width, feature size, and space of separation and overlap between layers including well, transistor, active area, poly layer, contact, via, and metal.
(iii) Mask design pattern: mask design is to ensure that the pattern can be exactly projected on wafer and locate the registration that mainly includes the shape of device feature, number of chip, and measurement registration.

During the fabrication process including manufacturing and test, six types of data would be automatically or semi-automatically recorded as follows:

(i) Production data: Production data contains basic information of every wafer in process including lot id, product name, process station, operation machine of the process station, operation time and date.
(ii) Metrology data: Metrology data contains the measurement data collected for specific quality characteristics (e.g. critical dimension, overlay, oxide thickness) to monitor and control process quality for a specific lot including lot id, measured parameter name of the product, measured parameter value of the product, specification of upper and lower limits of the product.
(iii) Equipment data: Equipment data contains the measurement data collected for a specific machine (e.g. temperature, pressure, etching rate) usually from pre-vention and maintenance including operation machine, measurement parameter of the machine, the measured value, measurement time and date, the specifi-cation of upper and lower limits of the machine.
(iv) Defect data: Defect data describes the defects of a specific product collected from the inspection equipments, failure analysis, scanning electron microscope (SEM), and signature analysis including lot id, product name, the defect layer, the number of defects in a layer, defect density, the number of defects in a wafer.
(v) WAT data: WAT contains test results including hundreds of parameters related to corresponding fabrication processes. In practice, a number of wafers will be

sampled from a lot and five points one the wafer are sampled to compare electric parameters with the standards to monitor the quality.

(vi) CP data: CP data contains the results of circuit probe tests of the dies in wafer sort including lot id, product name, and the die location of the wafer. CP test involves various functional tests for all the dies on each wafer. In particular, CP data consists of the CP bin summary data that is the sum of dies of a specific fail bin in a wafer that could be correlated with specific layer of wafer manufacturing process. The CP wafer bin map (WBM) that shows spatial patterns caused by specific manufacturing problems.

The amounts of collected data consist of useful patterns or causal relationship among design, manufacturing and test. For example, the problem of low CP yield may be related from particular manufacturing process which includes process station, process machine and production time. The failure of root cause could be identify by analyzing the CP data and production data. However, because the IC design house and wafer foundry concentrate on die design and manufacturing, respectively, the data between design and manufacturing has a gap that IC layout designs mainly focus on the area, speed, power and testability without considering the variation during wafer manufacturing.

Semiconductor fabrication facilities (fabs) can only maintain competitive advantages by effectively controlling process variation, fast yield ramp up, and quick response to yield excursion, especially when the complexity of the process and product increase rapidly. In particular, most of applications using various data mining technologies included root cause identification [8, 9], process improvement [10], defect pattern diagnosis [11], equipment backup control [12], cycle time prediction [13, 14], demand forecast [15, 16], and virtual metrology [17, 18]. Most applications are yield improvement for wafer manufacturing and test phase. The IC design phase has played a critical factor for yield improvement in nanometer. However, there are little research has been applied data mining methodologies for IC design phase. To bridge the gap between design and manufacturing, the input-output relationships of independent designable variables and dependent response variables should be identified in advance.

3 Framework of Design for Semiconductor Manufacturing

The proposed analytic framework of design for semiconductor manufacturing includes four phases: (i) problem definition and structuring, (ii) data preparation, (iii) model construction, and (iv) result evaluation and implementation as shown in Fig. 2.

The overall scheme considers the interactions among area, yield, variability, power, and speed have to be a trade-off during the whole IC design process and mask design process. In addition to above considerations, yield and cost for wafer manufacturing which is mainly correlated with area in layout design and mask design must be considered for competitiveness enhancement for mask producer and wafer fabrication.

During wafer manufacturing process, particular metrology parameters related to each manufacturing process are measured for ensuring high yield of the wafer.

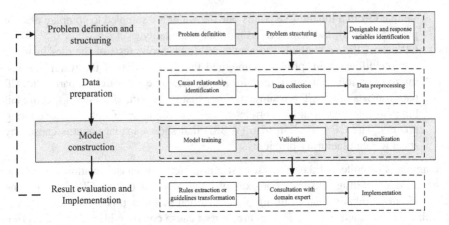

Fig. 2. Conceptual analytic framework of design for manufacturing

For example, overlay and critical dimension are measured in lithography process; etch rate and uniformity are measured in etch process. Although most process variation could be controlled and compensated via advanced manufacturing technology or advanced process control system in the wafer fabrication, systematic defect yield loss is significantly increasing as the feature size shrinking into nanometer scale. In particular, the design-based yield loss and lithography based yield loss which are parts of systematic losses become significant causes of low yield. Therefore, yield improvement should be embedded in IC design process to make the variation within acceptable tolerance.

To achieve the high throughput and yield, it is important to capture the information on how various IC geometric patterns of layouts influence the first important thing to do is circuit responses. The slight adjustment for the designable variables often results in the abrupt performance or unexpected variation. The better layout design pattern can lead to a direct or indirect basis of improvement. Since the nonlinear nature and the complexity increasing exponentially with advances in technology for layout design variables and manufacturing response, the above predictive model cannot be precisely formulated and predict the response exactly.

Preparing data in advance is an important and crucial step for enabling that the collected data can be input directly to further complex modeling for data analysis. The objectives of data preparation not only ensure data availability for analysis but also improve mining result effectiveness by improving data quality. In addition, data preparation also transforms the data value into correct format by the assumption of used data mining technique. The first is to identify the potential causal relationship between response and designable variables. The second is to design the experiment for collecting relevant data. Finally, collected data must be inspected and preprocessed to enhance effectiveness of further data modeling.

The modeling methods should be varied for different analytic purposes such as root cause identification and cycle time prediction. The IC design phase has played a critical factor for cost reduction and yield improvement for semiconductor manufacturing.

Because the involved response variables (dependent variables) and designable variables (independent variables) are continuous, the input-output relationship model can be formed into a regression model. In particular, multiple linear regression analysis is to construct a regression function and use the least square method for parameter estimation by minimizing the sum of squared errors that are the difference between actual response and predicted value. In general, the response variable y can be correlated with k designable variables x_k and the model is expressed as follows.

$$y = \beta_0 + \beta_1 x_1 + \beta_2 x_2 + \cdots + \beta_k x_k + \varepsilon \tag{1}$$

where the parameters β_i, $i = 0, 1, \cdots, k$, is called regression coefficients. Moreover, to consider the nonlinear relationship for the manufacturing response to the layout design variable, the effect of interaction term or high order term may be added into the model, for example, $x_1 x_2$ and x_1^2. However, the linear regression model is based on particular statistical assumptions and is not effective way of representing the function of whole nature relationship. On the other hand, the complicate relationship may be difficult for formulation. Therefore, neural network and rule-based methods are adequate for representing the response model.

Neural network is another technique for constructing the response model by forming an input-output network. Using neural network does not need assumptions such as independency among the variables and normality of data distribution. In particular, backpropagation (BP) network is a widely use of supervised learning for prediction or regression problem. Although neural network can deal with the complicate and high-order effect well, the model provides little information about the relationship about the data.

Regression tree, which is developed a tree structure by recursive partitioning as other decision tree, is used to derive IF-THEN rules for approximating the response model. The estimated responses in each leaf node are constant values. In addition, the IF-THEN rules can provide useful information for understanding the relationship between response variables and designable variables. Furthermore, model tree, another tree-based method called M5 [19], has been proposed to construct model for predicting continuous responses based on k predicted variables. Model tree differs from regression tree is that build piecewise linear model at the leaf node rather than a piecewise constant value. Model tree can generate smaller tree structure and represent more clearly than regression tree [19].

Generally, this study uses different tools for different situations of problems and data. No one method can be suited for all data and dominates others. The effectiveness of the constructed response model should be evaluated to see whether the relationship between layout designs and manufacturing responses can be accurately represented. Before implementation, the causal relation model should be discussed with domain expert for validation. The model can become a basis of constructing DFM guidelines including size, shape, space, or location of layout design for designers to improve manufacturability. Therefore, IC designer can adjust layout pattern for capturing better response of manufacturing performance under design specification.

4 Application to Layout Design of Chip Size

In the design phase, the designer typically chooses the chip size large enough to accommodate the micro-circuitry of ICs to ensure desired functionality. IC designer provides the pattern layout size information including chip size and chip shape for wafer fabrication. Given the fixed feature design, photolithography engineer in fabrication determines an optimal die placement positions with maximum gross die number and minimum shot number as the wafer exposure patterns. Moreover, the cycle time of wafer exposure has to be decreased for increasing more wafer per hour through reducing the shot number (# shot) of exposure machine. The utilization of mask field implies the efficiency of each shot as shown in Fig. 3. To find a better chip feature for exposing with few shot number per wafer without reduction of gross die number simultaneously.

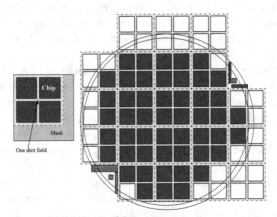

Fig. 3. Wafer exposure

To bridge the gap between chip size layout and wafer exposure, the behavior model of gross die and exposure shot is need to be constructed by various chip size. In practice, the optimal designs with maximum gross die number and minimum shot number are not always best while the modification of best chip layout design is difficult and time-consuming. In addition, the improved direction is difficult to find out because of the nonlinear relationship between gross die number per wafer and chip size. However, data mining methods can be used to not only extract models to explore the relationship between response variable and designable variables but also construct the behavior model for extracting the hidden and valuable information from the collected data during design and manufacturing.

Without losing generality, the data used in this empirical study has been systematically transformed before further analysis. First, the metric called mask-field-utilization weighted overall wafer effectiveness (MOWE) [20] is used as response variable to integrate the number of gross die and shot number. To construct the behavior mode for representing the relationship between MOWE and chip layout design size, there are 53561 data collected from the square region. The die length, die width and its corresponding gross die number per wafer and mask-field-utilization are recorded. Since the model tree, classification and regression tree (CART), regression analysis and back-propagation neural network (BPNN) can be used for both continuous response variable and predicted variable, therefore, the data is not required for transformation. Furthermore, the 60 % data are used for constructing the behavior model and 40 % data are used for model evaluation. In particular, the 60 % data are collected according the stratified sampling strategy for covering the whole two-dimensional pattern. In order to enhance the high prediction accuracy of model, thus the data intervals were cut strictly.

To evaluate the validity of used method, the M5 model tree was compared among CART and BPNN for representative of regression tree and neural network, respectively. Among methods were used the same training dataset for building response model and used the testing dataset for validating the performance. As shown in Table 1, the root mean squared error (RMSE) of M5 model tree was lower than CART and BPNN. In particular, the RMSE of BPNN was higher than tree-based model. In addition, the relative error of BPNN was larger than both M5 and CART. In particular, because the geometric characteristics of two-dimensional layout design, BPNN is difficult to perform the model fitness well while the response of MOWE is shape. Therefore, the M5 and CART which use recursive partitioning regression to extract the sharp boundary can be superior to BPNN.

Table 1. Comparison among M5 model tree, CART and BPNN by testing dataset

	Model tree (M5)	Regression tree (CART)	Neural Network (BPNN)
RMSE (%)	0.080	0.135	2.558
Relative error (%)	0.018	0.051	18.11

The metric called overall wafer effectiveness (OWE) and mask-field-utilization (MFU) are used to measure the number of gross die and shot number [20]. A generic product is used to illustration as shown in Table 2. To compare with the original chip size, the mask-filed-utilizations (MFU) of both are larger and only the number of gross die of pattern ii is better than original. In particular, comparing with Pattern i and Pattern ii results, it also shows that the higher MFU could result lower number of gross die. Furthermore, to consider the chip area is varied, the die area can be adjusted more

Table 2. Chip size adjustment with different condition

Scenario	#	Width	Length	Area	OWE (%)	MFU (%)
Original	–	8.922	8.9809	80.128	83.15	57.29
Area is fixed	i	12.620	6.3493	80.128	82.73	95.47
	ii	7.890	10.1557	80.128	83.37	84.84
Area is varied	iii	9.330	8.590	80.145	83.39	57.29
	iv	8.245	9.717	80.117	83.47	85.83
	v	8.820	9.073	80.024	83.69	57.22

extensively with some limitations and result in different OWE and MFU. Patterns iv can get the design with high OWE and MFU. Pattern iii and Pattern v both can capture high OWE, yet losing a scrap of MFU.

5 Conclusion

This study proposes an analytic framework of design for semiconductor manufacturing process. The proposed framework formulates the causal relationship between manufacturing response variables and designable variables into as prediction problem. Considering the problem, collected data quality, and result implementation, different kinds of data mining methods for constructing the response model should be applied. Through analyzing amount of data for layout design and manufacturing process, data mining can build an empirical model for IC designers to understand the relationship between the designable and response variables while the relationship is difficult to express via a physical formula. Furthermore, because of the shorter product life cycle, IC designer are under greater pressure to shorten time to market and provide a high-level product to maintain their competitive advantages in a changing and competitive environment. The extracted patterns or derived rules through the proposed framework can assist designer in validating the assumptions of design variables and become a basis of improvement direction.

Acknowledgments. This research is supported by National Science Council, Taiwan (MOST 103-2221-E-155-029-MY2).

References

1. Chiang, C., Kawa, J.: Design for manufacturability and yield for nano-scale CMOS. Springer, The Netherlands (2007)
2. White, K.P., Trybula, W.J., Athay, R.N.: Design for semiconductor manufacturing— perspective. IEEE Trans. Compon. Pack. Manuf. Technol. **20**(1), 58–72 (1997)
3. Gerez, S.: Algorithms for VLSI Design Automation. Wiley, New York (1999)
4. Wolf, W.: Modern VLSI Design: System-on-Chip Design, 3rd edn. Prentice-Hall PTR, London (2002)
5. International Technology Roadmap for Semiconductors: Design, International Technology Roadmap for Semiconductors 2005 edition (2005)
6. Weste, N., Eshraghian, K.: Principals of CMOS VLSI Design: A Systems Perspective. Addison-Wesley, Boston (1993)
7. Braha, D.: Data Mining for Design and Manufacturing: Methods and Application. Kluwer Academic publishers, Boston (2001)
8. Kusiak, A., Kurasek, C.: Data mining of printed-circuit board defects. IEEE Trans. Robot. Autom. **17**(2), 191–196 (2001)
9. Chien, C., Wang, W., Cheng, J.: Data mining for yield enhancement in semiconductor manufacturing and an empirical study. Expert Syst. Appl. **33**(1), 192–198 (2007)
10. Tsuda, H., Shirai, H.: Improvement of photolithography process by second generation data mining. IEEE Trans. Semicond. Manuf. **20**(3), 239–244 (2007)

11. Hsu, S., Chien, C.: Hybrid data mining approach for pattern extraction from wafer bin map to improve yield in semiconductor manufacturing. Int. J. Prod. Econ. **107**(1), 88–103 (2007)
12. Chien, C., Hsu, C.: A novel method for determining machine subgroups and backups with an empirical study for semiconductor manufacturing. J. Intell. Manuf. **17**(4), 429–440 (2006)
13. Kuo, C., Chien, C., Chen, J.: Manufacturing intelligence to exploit the value of production and tool data to reduce cycle time. IEEE Trans. Autom. Sci. Eng. **8**(1), 103–111 (2011)
14. Chien, C., Hsu, C., Hsiao, C.: Manufacturing intelligence to forecast and reduce semiconductor cycle time. J. Intell. Manuf. **23**(6), 2281–2294 (2012)
15. Çakanyildirim, M., Roundy, R.: SeDFAM: semiconductor demand forecast accuracy model. IIE Trans. **34**(5), 449–465 (2002)
16. Chien, C., Chen, Y., Peng, J.: Manufacturing intelligence for semiconductor demand forecast based on technology diffusion and product life cycle. Int. J. Prod. Econ. **128**(2), 496–509 (2010)
17. Hung, M., Lin, T., Cheng, F., Lin, R.: A novel virtual metrology scheme for predicting CVD thickness in semiconductor manufacturing. IEEE-ASME Trans. Mechatron. **12**(3), 308–316 (2007)
18. Lin, K., Hsu, C., Yu, H.: A virtual metrology approach for maintenance compensation to improve yield in semiconductor manufacturing. Int. J. Comput. Intell. Syst. **7**, 66–73 (2014)
19. Quinlan, J.R.: Learning with continuous classes. In: Proceedings of 5th Australian Joint Conference on Artificial Intelligence. World Scientific, Singapore, pp. 343–348 (1992)
20. Chien, C., Hsu, C., Chang, K.: Overall wafer effectiveness (OWE): a novel industry standard for semiconductor ecosystem as a whole. Comput. Ind. Eng. **65**(1), 117–127 (2013)

Workload and Delay Analysis in Manufacturing Process Using Process Mining

Minjeong Park[1], Minseok Song[1(✉)], Tae Hyun Baek[2],
SookYoung Son[2], Seung Jin Ha[2], and Sung Woo Cho[2]

[1] UNIST (Ulsan National Institute of Science and Technology),
Ulsan 689-798, Republic of Korea
{pmj9959,msong}@unist.ac.kr
[2] Hyundai Heavy Industries, Co, Ulsan 682-792, Republic of Korea
{bth,sson,haie,swcho00}@hhi.co.kr

Abstract. Process analysis is one of the important topics in manufacturing industry. Recently process mining has been applied to analyze manufacturing processes. In this paper we investigate the characteristics of event logs in make-to-order production and propose a method to analyze manufacturing processes in make-to-order production such as construction, shipbuilding and aviation by utilizing and extending existing process mining techniques. Among three major analysis perspectives in process mining such as process discovery, performance analysis, and conformance checking, this paper focuses on the performance analysis including workload analysis and delay analysis. To validate the proposed method, a case study with real data is conducted.

Keywords: Process mining · Manufacturing process analysis · Manufacturing execution system

1 Introduction

Manufacturing process refers to a process that is composed of sequential activities to transform raw materials into a finished product [23, 24]. To improve the quality of products and reduce production cost and time, many companies have tried to analyze complex manufacturing processes [4, 6]. Furthermore, many studies on manufacturing process analysis have been performed from several perspectives, such as process management [10], simulation [22], process modelling [7, 12], process performance analysis [8], fault detection [18], etc. Along with the methods mentioned above, process mining can be applied to analyze the manufacturing process. Process mining attempts to extract meaningful process-related information from event logs [13, 20, 21]. Manufacturing process analysis using process mining is used to understand current manufacturing processes by deriving process models, organizational models, social networks, etc. [14].

In this paper, we propose a way to analyze manufacturing processes in make-to-order industries. We investigate the characteristics of event logs in make-to-order

© Springer International Publishing Switzerland 2015
J. Bae et al. (Eds.): AP-BPM 2015, LNBIP 219, pp. 138–151, 2015.
DOI: 10.1007/978-3-319-19509-4_11

production and propose a method to analyze manufacturing processes in make-to-order production, such as construction, shipbuilding, and aviation by utilizing and extending existing process mining techniques. Major characteristics of make-to-order production are the existence of a plan of the manufacturing processes, several parallel activities, and different levels of process granularity. By investigating the characteristics, several issues in make-to-order production analysis are provided; these are performance analysis considering plans, conformance checking considering activity relations, frequency, resource, processing time and detailed parallel rules, process discovery that synthesizes cases of different levels, etc. Among these issues, we propose a way to analyze the performance of processes for make-to-order production in this paper. As a method for performance analysis, workload analysis and delay analysis using event logs are suggested. To validate the proposed method, a case study with real data is conducted.

The paper is organized as follows. Section 2 explains related work, including process mining and applications of process mining in the manufacturing industries. We present characteristics of make-to-order production processes in Sect. 3. In Sect. 4, we propose workload analysis and delay analysis. Section 5 conducts a case study to validate the proposed method and Sect. 6 concludes the paper with future work.

2   Related Work

The purpose of process mining is to discover, monitor, and improve actual processes from event logs recorded by Process-aware Information Systems (PAISs) such as enterprise resource planning (ERP), workflow management (WFM), customer relationship management (CRM), supply chain management (SCM), and product data management (PDM) systems [13, 20, 21]. Event logs are recorded by events in a consecutive order, and each event has an activity related to a case. In addition, the event can have a timestamp and resource [21]. Process mining is composed of discovery, conformance checking, and enhancement. Discovery is to derive models from event logs. The alpha algorithm, heuristic mining, fuzzy mining, etc. are examples of discovery techniques. Conformance checking is to compare a given model with corresponding event logs, and enhancement is to extend or improve a model using observed behaviors [16, 19]. Likewise, process mining has many techniques applicable for analyzing processes. Therefore, process mining has been applied in many domains, such as healthcare, service, logistics, public administration, manufacturing, and so forth. For example, Mans et al. provided insights for hospital processes by applying various process mining techniques in a control-flow perspective, organizational perspective, and performance perspective [11]. Weerdt et al. proposed a framework for actual process analysis in a multifaceted financial service industry using process mining [3]. In addition, van der Aalst et al. applied process mining to the Dutch National Public Works Department managing roads and bridges to analyze invoice processes [21], Bozkaya et al. suggested a methodology for process diagnostic based on process mining and applied it to processes of government agencies [2], and Jeon et al. proposed a conceptual framework for identifying causes of inefficiency in port logistics using process mining [9].

There have been few studies in the manufacturing industries, except for Rozinat et al. who studied the applicability of process mining in ASML, a leading wafer scanner manufacturing company, and provided proposals for improvement [13]. In addition to a shortage of application cases, previous studies are difficult to apply to the make-to-order industries owing to the characteristics of their processes. Thus, this paper proposes a method of manufacturing process analysis applying process mining considering the major characteristics of the processes.

3   Characteristics of Make-to-Order Manufacturing Processes

In this section, the characteristics of make-to-order manufacturing processes are discussed. First, there exists a detailed plan for manufacturing processes. In the case of the make-to-order manufacturing industry, it is very important strictly to follow optimized and planned processes, because meeting a deadline for the clients' orders is considered as the top priority [12]. Therefore, manufacturing process analysis needs to compare a plan and an actual production (e.g., how much the actual process is being conducted as planned). Moreover, it is important to compare resources and processing time as well as the order of activities between the plan and the result.

Another characteristic of make-to-order manufacturing processes is that the manufacturing processes have several parallel activities. Because the size of the products in the make-to-order manufacturing industry is usually large, several activities happen simultaneously. Furthermore, they sometimes strictly define the relationship between event types, such as start and end between the parallel activities. For example, if there are two activities in parallel, a rule can be that an activity must start before the other starts and both activities have to end at the same time. Another rule can be that both activities have to start simultaneously and an activity must end before the other ends.

The third characteristic is that manufacturing process models usually have different levels of granularity [5]. In other words, an overall manufacturing process consists of many different subprocesses. For example, in a project for an offshore plant production, an offshore plant consists of some modules and each module consists of a number of blocks. For each block, there is a corresponding manufacturing process that is composed of various activities. These blocks build up to become a finished module. The modules also have their individual manufacturing processes for completion after combining blocks. Because the overall process has an intricate network among subprocesses and unit activities, finding a proper case is difficult in applying process mining.

In this paper, we propose a method to analyze performance of make-to-order manufacturing processes. Considering one of the major characteristics mentioned above, a different performance analysis from the traditional analysis is required. The existing performance analysis in manufacturing processes has been measuring only the results of the actual processes being conducted to compare with the criteria the business sets as its goal [10]. Nonetheless, in the case of make-to-order manufacturing industries, there are not only the actual manufacturing processes but the planned manufacturing processes. Thus, it is required to conduct manufacturing process analysis on both the actual and planned processes to identify how well the actual processes are being executed as planned. The planned processes are optimized considering the costs, efforts, and time. This is why the more the actual processes conform to the planned processes, the better the result becomes. Hence, our research tries to determine indicators to measure quantitatively the performance based on the comparison of how the actual processes deviate from the planned processes.

4 Analysis for Measuring Performance on Manufacturing Processes

In order to conduct performance analysis of manufacturing processes, we first define an event and an event log. As noted in Sect. 3, the structure of event logs for make-to-order processes is different in that it includes attributes of result as well as plan. An event and an event log are defined as follows.

Definition 1. (Event of manufacturing processes) Let A be a set of activities (e.g., tasks), R_{plan} and R_{res} be a set of resources (e.g., departments or workers) for plan and result, ST_{plan} and ST_{res} be a set of start timestamps for plan and result, CT_{plan} and CT_{res} be a set of complete timestamps for plan and result respectively. $E = A \times R_{plan} \times R_{res} \times ST_{plan} \times ST_{res} \times CT_{plan} \times CT_{res}$ is a set of events and π_n is the value of attribute n for an event $e = \{a, rp, rr, stp, str, ctp, ctr\}$.

Definition 2. (Event log of manufacturing processes) Let C be a set of events corresponding each case. $L \in \mathcal{B}(C)$ is an event log and $\mathcal{B}(C)$ means all bags over C.

Table 1. An example of events for manufacturing processes

CaseID	EventID	A	Rplan	Rres	STplan	CTplan	STres	CTres
BlockA	01	Act_1	R_a	R_a	2015-01-25	2015-03-01	2015-01-26	2015-03-01
	02	Act_2	R_b	R_b	2015-02-04	2015-02-15	2015-02-04	2015-02-14
BlockB	03	Act_1	R_a	R_a	2015-02-06	2015-02-14	2015-02-06	2015-02-15
	04	Act_2	R_b	R_b	2015-02-27	2015-02-27	2015-02-27	–
	05	Act_3	R_c	R_d	2015-03-26	2015-04-20	–	–

Table 1 shows an example of events. From the table, there are five events (E = {01, 02, 03, 04, 05}). There are three activities (A = {Act_1, Act_2, Act_3}) and four resources (R = {R_a, R_b, R_c, R_d}). Start timestamps for plan and result (ST$_{plan}$ = {2015-01-25, 2015-02-04, 2015-02-06, 2015-02-27, 2015-03-26}, ST$_{res}$ = {2015-01-26, 2015-02-04, 2015-02-06, 2015-02-27}) and complete timestamps for plan and result (CT$_{plan}$ = {2015-03-01, 2015-02-15, 2015-02-14, 2015-02-27, 2015-04-20}, CT$_{res}$ = {2015-03-01, 2015-02-14, 2015-02-15}) are also shown in the table.

We conduct workload analysis and delay analysis as performance analysis. Workload analysis aims to find the degree of workload on resources by measuring and analyzing the number of activities for the planned and the result using event logs. Delay analysis attempts to understand the degree of delay on activities or resources.

4.1 Workload Analysis

Workload analysis measures the number of activities performed by each resource in a certain period. Workload analysis shows the number of started activities, completed activities, and activities in progress. Using activity frequencies, workload analysis provides information on how much the workload of each resource is. It gives insights on how to control the manufacturing processes based on the number of activities in progress. The definitions of the number of start activities, end activities, and activities in progress in a unit period are as follows.

Definition 3. (The number of activities ()) For $e = (e_0, e_1, e_2, \ldots) \in L$, $a_j \in A$, $rp_k \in R_{plan}$, $rr_l \in R_{res}$, $stp_m \in ST_{plan}$, $str_n \in ST_{res}$, $ctp_o \in CT_{plan}$ and $ctr_p \in CT_{res}$, the number of activities:

- The number of started activities for plan (a_j, rp_k, stp_m)

$$= \sum_{0 \le e < |L|} \sum_{0 \le i < |e|} \begin{cases} 1 & \begin{aligned} &\text{if } \pi_a(e_i) = a_j \\ &\wedge \pi_{rp}(e_i) = rp_k \\ &\wedge \pi_{stp}(e_i) = stp_m \end{aligned} \\ 0 & \text{otherwise} \end{cases} \tag{1}$$

- The number of completed activities for plan (a_j, rp_k, ctp_o)

$$= \sum_{0 \le e < |L|} \sum_{0 \le i < |e|} \begin{cases} 1 & \begin{aligned} &\text{if } \pi_a(e_i) = a_j \\ &\wedge \pi_{rp}(e_i) = rp_k \\ &\wedge \pi_{ctp}(e_i) = ctp_o \end{aligned} \\ 0 & \text{otherwise} \end{cases} \tag{2}$$

- The number of started activities for result (a_j, rr_l, str_n)

$$= \sum_{0 \le e < |L|} \sum_{0 \le i < |e|} \begin{cases} 1 & \begin{aligned} &\text{if } \pi_a(e_i) = a_j \\ &\wedge \pi_{rr}(e_i) = rp_k \\ &\wedge \pi_{str}(e_i) = stp_n \end{aligned} \\ 0 & \text{otherwise} \end{cases} \tag{3}$$

- The number of completed activities for result (a_j, rr_l, ctr_p)

$$= \sum_{0 \leq e < |L|} \sum_{0 \leq i < |e|} \begin{cases} 1 & \begin{aligned} &\text{if } \pi_a(e_i) = a_j \\ &\wedge \pi_{rr}(e_i) = rp_k \\ &\wedge \pi_{ctr}(e_i) = ctp_p \end{aligned} \\ 0 & \text{otherwise} \end{cases} \tag{4}$$

- The number of activities in progress for plan $(a_j, rp_k, stp_m, ctp_o)$

$$= \sum_{\mathrm{Min}(a_j, rp_k) \leq t_l} \text{The number of started activities for plan } (a_j, rp_k, stp_m)$$

$$- \sum_{\mathrm{Min}(a_j, rp_k) \leq t_l} \text{The number of completed activities for plan } (a_j, rp_k, ctp_o) \tag{5}$$

where $\mathrm{Min}(a_j, rp_k) = \min(\forall \pi_t \{e_i | \pi_a(e_i) = a_j \wedge \pi_{rp}(e_i) = rp_k)$.

- The number of activities in progress for result $(a_j, rr_l, str_n, ctr_p)$

$$= \sum_{\mathrm{Min}(a_j, rr_i) \leq t_l} \text{The number of started activities for result } (a_j, rr_l, str_o)$$

$$- \sum_{\mathrm{Min}(a_j, rr_l) \leq t_l} \text{The number of completed activities for result } (a_j, rr_l, ctr_p) \tag{6}$$

where $\mathrm{Min}(a_j, rr_l) = \min(\forall \pi_t \{e_i | \pi_a(e_i) = a_j \wedge \pi_{rr}(e_i) = rr_l)$.

The visualization based on workload analysis is shown in Fig. 1. The figure shows the number of started activities, completed activities, and activities in progress. The blue line represents the number of started activities. The green line and the red line represent the number of completed activities and activities in progress, respectively. The red line shows that the number of started activities steadily rose until early June, then decreased afterwards.

The workload analysis can be performed according to each resource or each activity. In the case of workload analysis based on resources, it is possible to analyze the number of activities of a particular resource in a certain period. The workload analysis helps to find where the overload points are and which resource has piled the activities in progress. Likewise, in the case of workload analysis based on activities, it provides information on how much of the workload is assigned to a particular activity.

Comparison for workload analysis between the planned and the actual proceeds as follows. First, workload analysis of the planned and the actual are performed individually. Based on the results, we calculate the statistical measures for the activities in progress and then compare them. The statistical measures include average, median, maximum, and minimum values. If the measures of the actual are higher than the planned values, we need to compare the values of the started and completed activities as well. Through this comparison, we can detect problems such as where the activities

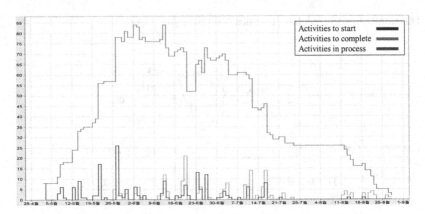

Fig. 1. An example of workload analysis (Color figure online)

are being overloaded and where the activities are not being completed in time. In addition, it is possible to understand the current state of the manufacturing processes and the diagnosis of problems through the workload analysis.

4.2 Delay Analysis

In make-to-order industries, delay is one of the factors causing the most common cost problems [1]. Delay analysis is performed to identify delayed activities and resources causing delays and to calculate the extent of delays. We define two measures, such as the delay based on completion date of an activity and the delay based on processing time. The delay is usually defined as an activity being completed later than the planned completion date [17]. Definition 4 shows the delay based on completion date.

Definition 4. (Completion date-based $Delay_{time}$ ()) For $a_q \in A$ and $e = (e_0, e_1, e_2, ...) \in L$, compeletion date-based $Delay_{time}$:

$$Completion\ date-based\ Delay_{time}(e_i, a_q) = \pi_{ctr}(e_i) - \pi_{ctp}(e_i)$$
$$where\ \pi_{ctr}(e_i) \neq null \wedge \pi_a(e_i) = a_q$$

The calculation of the delay based on the completion date is useful when meeting deadline is the most important. However, it does not provide sufficient information, because the delay on completion date might be caused by a late start. That is, if a preceding activity is delayed, then following activities seem to be delayed though processing times of activities are not delayed at all. Therefore, the delay based on the processing time has to be measured as the second method. The calculation is measured by how long the actual processing time of an activity is compared with the planned time.

Definition 5. (Processing time-based Delay$_{time}$ ()) For $e_i = (e_0, e_1, e_2,...) \in L$ and $a_q \in A$, processing time-based Delay$_{time}$:

Processing time $-$ based Delay$_{time}(e_i, a_q)$

$$= \left(\pi_{ctr}(e_i) - \pi_{str}(e_i)\right) - \left(\pi_{ctp}(e_i) - \pi_{stp}(e_i)\right) where \; \pi_{ctr}(e_i) \neq null \wedge \pi_a(e_i) = a_q$$

In conducting delay analysis, we first calculate the delays based on completion date and processing time. Then, we set the criteria for dividing the extent of the delays from discussion with the domain experts. Based on the results, we derive the ratios for each groups that is divided by the criteria. By doing so, activities that are frequently delayed can be detected. Likewise, it is possible to conduct the delay analysis from the resource perspective as well.

5 Case Study

In this section, we discuss a case study performed to validate the proposed method. The event logs are the offshore plant manufacturing process provided by Hyundai Heavy Industries Co., one of the major shipbuilding companies in the world. The case study is subdivided into data preparation, analysis, and discussion steps. In the data preparation, we collected the event logs from the company and performed the data preprocessing. Next, in the analysis, the performance analysis was conducted. Finally, the overall results are discussed, including comments from the domain experts in the discussion part.

5.1 Data Preparation

The event logs were collected from the information system in the company. This contains event logs for four ongoing projects. The attributes of the collected data are ProjectID, ModuleID, BlockID, ActivityCode, ActivityName, planned started and completed dates, actual started and completed dates, DepartmentCode, and Department-Name. Among the attributes, we used BlockID as a case, ActivityCode as an activity, and DepartmentCode as a resource. The started and completed dates for plan and started and completed dates for result are utilized as timestamps.

Data preprocessing was performed because the raw data are incomplete, inconsistent, and contain noise. In the data preprocessing, data cleansing, data transformation, and data conversion were conducted. Figure 2 shows the overall events of the logs using a dotted chart [15]. In the chart, we recognized events that have null in the resource attribute and filled in the proper values using the existing information.

Furthermore, we found that a start date of an activity was incorrect. Thus, data cleansing was done by domain experts. Finally, the data were converted into data types suitable for process mining (i.e., MXML, XES). As the result of the data preprocessing, the event logs were generated as follows: approximately 700 blocks, 1400 activities, 200 activity types, and 12 departments.

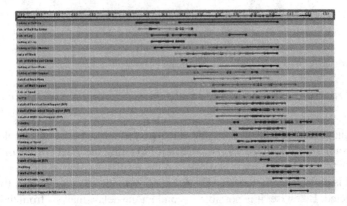

Fig. 2. An example of dotted chart results

5.2 Workload Analysis

We conducted workload analysis to understand the degree of workload of each department and each activity using the event logs.

Figure 3 is a result of the workload analysis on Department_a. The upper graph in Fig. 3 visualizes the workload analysis result for the planned and the graph below is for the actual. According to the planned, there are 21 concurrent activities on average. However, the actual number of on-going activities is 39 on average. Furthermore, in the case of the maximum number of activities, it is 64 on the planned while the actual is 95.

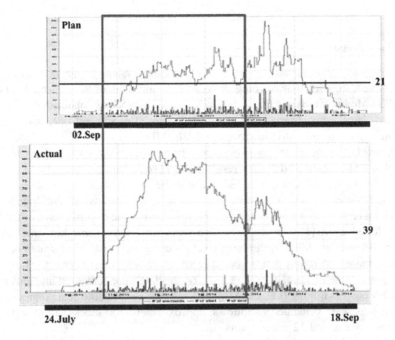

Fig. 3. Workload analysis results: department a

This shows that there were more concurrent on-going activities in practice than as planned. A problem regarding the overload can be found through additional comparison between the number of started and completed activities. The problem was caused by the completed activities that did not terminate in time as the planned. This led to an increase in the number of on-going activities. Even though the department started the activities 40 days earlier than they were planned, the activities were delayed and completed 20 days later. Figure 4 shows the workload analysis on the activities performed by Department_a. Among the five activities, Activity 2, Activity 3, and Activity 5 have the same problem shown in Fig. 3.

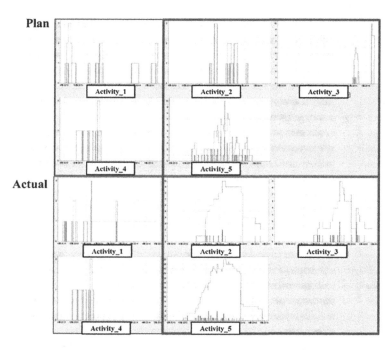

Fig. 4. Workload analysis results: activities of department a

5.3 Delay Analysis

We performed the delay analysis to determine the degree of the delays on each activity and each department using event logs. There are two criteria when measuring the delays, completion date-based and processing time-based. As shown in Fig. 5, the delay analysis was carried out for the top 10 most frequent activities out of 200.

Seven criteria were defined considering the extent of the delays. For instance, in Fig. 5(a), activities with no delay were classified into Criterion 1. Activities with less than five days of delay were classified into Criterion 2. Similarly, we classified activities with less than 10 days of delay into Criterion 3, less than 40 days of delay into Criterion 4, less than 50 days of delay into Criterion 5, less than 100 days of delay into Criterion 6, and 100 days and more of delay into Criterion 7. In the same way, we classified each activity by the processing time in Fig. 5(b). Afterwards, we calculated the frequency ratios for each criterion and visualized them in Fig. 5. In Fig. 5(a), we can identify that Activity_26 and Activity_27 were performed 90 % of the time with no delay. In contrast, Activity_93 and Activity_92 were mostly delayed. To understand the number of delays based on processing time, Fig. 5(b) is shown. Activity_93, which was delayed over 90 % of the time in Fig. 5(a), was completed 65 % without any delay in Fig. 5(b). The main reason why Activity_93 was more delayed in the completion date-based scheme rather than the processing time-based scheme is the actual started date itself was delayed from the planned date.

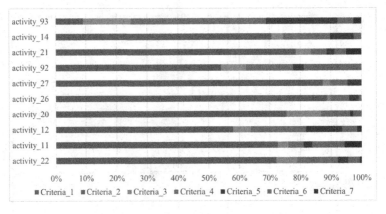

(a)   Completion date based

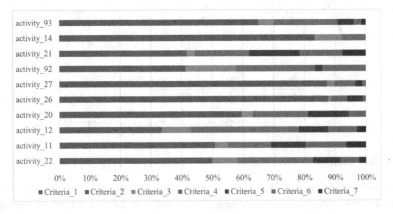

(b)   Processing time-based

Fig. 5. Delay analysis results: Activity perspective

5.4   Discussion

The previous section showed that performance analysis including workload analysis and delay analysis can be applied to real data. To evaluate the suggested method, we had several meetings with domain experts. Through the discussions, we obtained feedback on the results as follows.

A department located outside the factory can cause delays or overloads. In the results of workload analysis and delay analysis, Department_a had problems. According to the results of workload analysis, there were more activities in progress in actual processes than in planned processes in Department_a. In addition, we found that approximately 60 % of activities performed by Department_a were delayed from the results of processing time-based delay analysis. Through the discussion, we discovered that Department_a was located outside the factory, which makes it difficult to control and manage compared with other departments.

In addition, we found that some reworked activities were delayed. If a product with defects is found, reworks of activities are unavoidable to repair it. Once an activity is reworked, it typically takes a few more weeks compared with the planned, thus reworked activities may cause delays. For example, in the case of painting activities, they had been delayed by frequent reworks because the activities have a high failure rate.

Besides, there were other factors to bring about delays and overloads. Activities performed in the rainy season were more frequently delayed than activities in other seasons. About 37 % of delayed activities were performed in June to July, when the rainy season starts in Korea. In addition, improper planning of activities may cause overloads because low-priority activities would be delayed to deal with unexpected conditions. For example, if the number of resources is significantly changed from the plan or if raw materials do not arrive in time, the activity would not be completed at the planned date and the number of activities in progress would pile up.

We introduced performance analyses to determine which activities or departments have problems based on the results. Our study has an advantage in understanding the current situation by comparing the actual processes with the planned processes. On the other hand, it is difficult to identify exact causes of problems through the results. This limitation can be overcome by discussion with domain experts, as presented above. In addition, although most of the other analyses have the same issue, the results of the suggested performance analyses are highly dependent on the quality of data because event logs used in this research consist of planned data and actual data. For that reason, this study needs well-structured data from MES to obtain meaningful information from the obtained results.

6 Conclusion

This paper provided characteristics of manufacturing processes in make-to-order production and proposed a method for performance analysis including workload analysis and delay analysis. Workload analysis, which analyzes the degree of workload, and delay analysis, which examines the degree of delay in the activity and the resource perspectives, were proposed. In addition, a case study was conducted to validate the proposed method. In the case study, we were able to understand the current situation and detect the causes of the problems by comparing the planned with the actual processes. As future work, we plan to analyze a method to make a plan for precisely processing time of activities considering the level of difficulty of an activity measured by the results of performance analysis. At present, it is difficult to predict the processing time of each activity because even if the activities are the same, the processing time of an activity varies according to the level of difficulty. Therefore, reflecting the level of difficulty in conducting activities can be a solution to design a proper plan. In addition, we intend to develop an indicator that provides the degree of how much actual processes deviate from planned processes. Although conformance checking detects and quantifies inconsistency between event logs and models, it needs to be extended to deal with the manufacturing processes in make-to-order production. After discussion with the domain experts, we determined that not only activity relation and frequency but also resource, processing time, and detailed parallel relations should be considered when analyzing the conformance checking. Hence, we propose to study comprehensive conformance checking analysis for manufacturing processes in make-to-order industries using process mining.

Acknowledgments. This research was supported by the Basic Science Research Program through the National Research Foundation of Korea (NRF) funded by the Ministry of Education, Science and Technology (No. 2011-0010561).

References

1. Alkass, S., Mazerolle, M., Harris, F.: Construction delay analysis techniques. Constr. Manag. Econ. **14**, 375–394 (1996)
2. Bozkaya, M., Gabriels, J., Werf, J.: Process diagnostics: a method based on process mining. In: International Conference on Information, Process, and Knowledge Management, eKNOW 2009, pp. 22–27. IEEE (2009)
3. De Weerdt, J., Schupp, A., Vanderloock, A., Baesens, B.: Process mining for the multi-faceted analysis of business processes—a case study in a financial services organization. Comput. Ind. **64**, 57–67 (2013)
4. Edgar, T.F., Butler, S.W., Campbell, W.J., Pfeiffer, C., Bode, C., Hwang, S.B., Balakrishnan, K., Hahn, J.: Automatic control in microelectronics manufacturing: practices, challenges, and possibilities. Automatica **36**, 1567–1603 (2000)
5. ElMaraghy, H.A.: Reconfigurable process plans for responsive manufacturing systems. In: Cunha, P.F., Maropoulos, P.G. (eds.) Digital Enterprise Technology. Perspectives and Future Challenges, pp. 35–44. Springer, Heidelberg (2007)
6. Harding, J., Shahbaz, M., Kusiak, A.: Data mining in manufacturing: a review. J. Manuf. Sci. Eng. **128**, 969–976 (2006)
7. Hernandez-Matias, J.-C., Vizán, A., Hidalgo, A., Ríos, J.: Evaluation of techniques for manufacturing process analysis. J. Intell. Manuf. **17**, 571–583 (2006)
8. Hornix, P.T.: Performance analysis of business processes through process mining. Master's thesis, Eindhoven University of Technology (2007)
9. Jeon, D., Yahya, B., Bae, H., Song, M., Sul, S., Sutrisnowati, R.: Conceptual framework for container-handling process analytics. ICIC Express Lett. **7**, 1919–1924 (2013)
10. Lin, H., Fan, Y., Newman, S.T.: Manufacturing process analysis with support of workflow modelling and simulation. Int. J. Prod. Res. **47**, 1773–1790 (2009)

11. Mans, R.S., Schonenberg, M.H., Song, M., van der Aalst, W.M.P., Bakker, P.J.M.: Application of process mining in healthcare – a case study in a Dutch hospital. In: Fred, A., Filipe, J., Gamboa, H. (eds.) BIOSTEC 2008. CCIS, vol. 25, pp. 425–438. Springer, Heidelberg (2009)

12. Mohamed, N., Khan, M.: Decomposition of manufacturing processes: a review. Int. J. Automot. Mech. Eng. **5**, 545–560 (2012)

13. Rozinat, A., de Jong, I.S., Gunther, C., van der Aalst, W.M.P.: Process mining applied to the test process of wafer scanners in ASML. IEEE Trans. Syst. Man Cybern C Appl. Rev. **39**, 474–479 (2009)

14. Son, S., Yahya, B., Song, M., Choi, S., Hyeon, J., Lee, B., Jang, Y., Sung, N.: Process mining for manufacturing process analysis: a case study. In: Proceeding of 2nd Asia Pacific Conference on Business Process Management, Brisbane, Australia (2014)

15. Song, M., van der Aalst, W.M.P.: Supporting process mining by showing events at a glance. In: Proceedings of the 17th Annual Workshop on Information Technologies and Systems (WITS), pp. 139–145 (2007)

16. Song, M., van der Aalst, W.M.P.: Towards comprehensive support for organizational mining. Decis. Support Syst. **46**, 300–317 (2008)

17. Talluri, S.: A benchmarking method for business-process reengineering and improvement. Int. J. Flex. Manuf. Syst. **12**, 291–304 (2000)

18. Tsung, F., Zhou, Z., Jiang, W.: Applying manufacturing batch techniques to fraud detection with incomplete customer information. IIE Trans. **39**, 671–680 (2007)

19. van der Aalst, W.M.P., Adriansyah, A., van Dongen, B.F.: Replaying history on process models for conformance checking and performance analysis. Wiley Interdiscip. Rev. Data Min. Knowl. Discov. **2**, 182–192 (2012)

20. van der Aalst, W.M.P., Reijers, H.A., Weijters, A.J., van Dongen, B.F., de Alves Medeiros, A., Song, M., Verbeek, H.: Business process mining: an industrial application. Inf. Syst. **32**, 713–732 (2007)

21. van der Aalst, W.M.P., Schonenberg, M.H., Song, M.: Time prediction based on process mining. Inf. Syst. **36**, 450–475 (2011)

22. Vignat, F., Villeneuve, F.: Simulation of the manufacturing process, generation of a model of the manufactured parts. digital enterprise technology. In: Cunha, P.F., Maropoulos, P.G. (eds.) Digital Enterprise Technology, pp. 542–552. Springer, Heidelberg (2007)

23. Wang, Z., Du, P., Yu, Y.: An intelligent modeling and analysis method of manufacturing process using the first-order predicate logic. Comput. Ind. Eng. **56**, 1559–1565 (2009)

24. Wen, L., Tuffley, D.: Formalizing manufacturing process modeling using composition trees. Adv. Mat. Res. **399**, 1852–1855 (2012)

Emerging Topics in BPM

Social-Network-Based Personal Processes

Seyed Alireza Hajimirsadeghi$^{(\boxtimes)}$, Hye-Young Paik, and John Shepherd

University of New South Wales, Sydney, NSW 2052, Australia
{seyedh,hpaik,jas}@cse.unsw.edu.au

Abstract. In this paper, we propose *Processbook*, a social-network-based management system for *personal processes* (ad hoc processes carried out to achieve a personal goal). A simple modelling interface is introduced based on ToDoLists to help users plan towards their goals. We describe how the system can capture a user's experience in managing their ToDoList and the associated personal process, how this information can be shared with other users, and how the system can use this information to recommend process strategies. We exemplify the approach by a sample administrative process inside University of New South Wales and report some preliminary evaluations of the proposed recommendation algorithms.

Keywords: Personal process management · Task management systems · Process knowledge · Process recommendation · Social networks

1 Introduction

In modern society, we are frequently required to perform administrative or business processes in order to achieve our goals. Examples could be simple processes such as booking a theatre ticket, or more complicated and long lasting ones such as planning for immigration.

While the last decade has seen many of these individual processes codified via online services, there remain significant problems in discovering and integrating the very many tasks that have not yet been codified. An important aspect of the problem is that processes frequently span organisational boundaries and there are few mechanisms to carry information and outcomes from processes in one organisation to those in the next organisation. Another major factor is that it is sometimes difficult to identify precisely which organisations and which processes within those organisations are required to accomplish a stated goal.

A *personal process* is made up of a number of tasks which needs to be carried out in order to achieve a goal; a task may be as simple as one individual activity or may be as complex as a complete business process. Personal processes often require an integration of so-called "long tail" business processes from one or more organisations. The knowledge of such integration and consequently the knowledge of achieving a goal is referred to as *process knowledge* in this paper.

Our goal is to provide support for individuals to manage their personal processes. One possible approach for this would be to convert all personal

© Springer International Publishing Switzerland 2015
J. Bae et al. (Eds.): AP-BPM 2015, LNBIP 219, pp. 155–169, 2015.
DOI: 10.1007/978-3-319-19509-4_12

processes into codified business processes, but this is neither plausible nor cost-effective [7]. Instead, we aim to assist users in discovering the tasks they have to do to reach their goals, the order they need to do the tasks and the sets of constraints and rules they should follow in doing those tasks. This is accomplished in a context where other users have already successfully carried out the processes and have recorded the method by which they did so. Our system to support this (*Processbook*) adopts a social-based approach, aimed at non-expert users, who describe their personal processes via *ToDoLists*. *Processbook* collects, manages, merges and shares process descriptions and allows users to follow other users who are carrying out similar tasks and can recommend future steps in a given process based on how others users previously achieved the same goal.

In Sect. 2, we elaborate the problem area with examples. Section 3 describes related work in the space of personal processes and using social software in business process management. Section 4 presents the details of *Processbook* and explains how process knowledge is captured and then recommended to users. In Sect. 5 we investigate the implemented tool with an example in academic domain. Finally we conclude the paper in Sect. 6 with future work.

2   Personal Process Management: A Motivating Scenario

In this section, we consider the problem of carrying out personal processes via an example: Ali, a student from a non English-speaking country, wishes to study for a PhD in one of the top 8 Australian universities (known as Go8). Ali has two primary objectives: find a university that would accept him and maximise the amount of funding to assist his study. Additional constraints and preferences might include: a PhD topic in the service oriented computing area, a PhD program commencing after July 2013, etc.

Figure 1a illustrates how Ali plans to reach his goals and what sources he utilises for his purpose. He tries to identify universities that satisfy his constraints by asking friends or by searching on the Web, collects and collates information about the entry requirements and scholarship availability for each university from its official web site. He might also join relevant communities on popular social networks such as facebook and twitter to keep up with the latest news and updates about the institutions he is dealing with.

In carrying out the above, questions would arise at each stage for Ali. For example, the web site at some university might specify that a student needs to provide an undergraduate transcript and English proficiency test results, but might not mention the kind of visa that the student requires or how to obtain such a visa. Other typical questions that might arise are: what step should I take next, what do I do at each step, which organisation should I deal with, etc. To find the answers to such questions, Ali would seek answers from his friends, experts, social networks or other data available in online forums, blogs, etc. Ali uses a to-do list approach to organise his findings and to write down the tasks he has to do. He may do so by simply writing on paper, using computer-aided task management tools or registering in an online task management website where he is offered more gadgets to organise his tasks.

(a) Example of a Personal Process Management

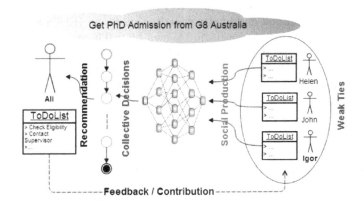

(b) Personal Process Management in *Processbook*

Fig. 1. From personal involvement to active social participation

Getting advice from someone experienced with the specific process would be extremely useful, but finding such an expert might be difficult. A more effective approach might be to have the process information available online, and have a system that understands both the process information and your personal situation (in terms of progress through the process), and can offer sufficient information to enable you to determine how to proceed. In practice, a number of difficult issues need to be dealt with before such a system can be realised:

Invalid, Incomplete or Inconsistent Data: We may be faced with untrusted sources of information, or conflicting items of information, or may be given out-of-date information. Sometimes, we simply do not know certain parts of the process. In other cases, there may be hidden (or ignored) pieces of information. For example, Middle Eastern students may face a wait of up to three months in applying for a Australian student visa.

Individuals as Process Integrators: Individuals are responsible to collect all the relevant data and integrate it effectively while planning their goals. This is a challenging task as the process domain is usually new to individuals and they do not have an overview of the process when progressing step by step. Heterogeneous data sources and data formats, numerous data dependencies over multiple organisations and constantly changing policies and workflows makes it even more difficult to play the role of process integrator.

Inability to Predict Task Effects: Sometimes it is difficult to know what kind of effect accomplishing a specific task will have on the process as a whole. For example, while either of the IELTS and TOEFL English competency tests are accepted world wide, it is better to have IELTS scores if applying for Australian universities because they are better regarded.

Isolated Individuals: Although people join communities on social networks and discuss issues with peers in forums and e-how sites, they are ultimately progressing in an isolated manner. In Fig. 1a, Igor is a member of Go8 communities on facebook and is also contributing on an IELTS message board, while reflecting his experiences in his own blog. There is no established way for Ali to be aware of all Igor's contributions on the process, to contact him or to learn from his progress. People with similar goals and interests may not be able to find each other easily and each individual's progress is not necessarily recorded for future reuse.

In *Processbook*, we consider merging social networks principles with process management basics to overcome the above issues. As Fig. 1b illustrates:

- A goal-based community is established to remove the isolation barrier and help individuals find peers with similar interests/goals more easily.
- Users' knowledge and experience is captured unobtrusively while they are planning for their goal via a simple modelling interface that requires no prior process modelling knowledge.
- Individuals' process knowledge is merged to produce a general view of the process.
- Users' votes and comments are added to the socially produced process knowledge to minimize the adverse effect of invalid, inconsistent and incomplete data.

3   Related Work

Personal process management has so far received limited attention from the academic research community. A vision statement can be found on the blog posted by Michael Rosemann[1]. Two possible implementations of personal process management are discussed in [7] and [1].Reference [7] mainly focuses on sequential and conditional constraints by introducing a formal personal process modelling language. The proposal in [1] is based on parallel executions, tries to simplify BPM techniques and pays attention to the role of social aspects of the process management such as sharing and assigning tasks. Both works remain at preliminary level and are yet to realise any significant improvement over personal process management.

---

[1] http://www.michaelrosemann.com/uncategorized/113/.

On the other side, a large number of commercial online tools exist for personal task management. These tools are end-user oriented and provide a plethora of features including task creation, sharing, social network integration, notification, etc. However, as [1] notes, none of them is concerned about the "process"concept; they do not embrace the practices of BPM, thus losing many beneficial aspects of structuring dependencies and constraints between tasks.

In terms of requiring flexibility and agility, personal process management is closely related to agile BPM. The most notable defect of classical BPM is the "model-reality divide", the distance between abstract process models and the processes executed in practice. To overcome this defect, [2] states that agile BPM not only requires changes to the BPM life cycle, but also a paradigmatic change. This paradigmatic change can be obtained by applying social software features into business process management.

There have been some attempts in recent years to accommodate social features in the BPM environment. Most notably a recommendation-based process modelling support system with social features in [4], a modelling and execution tool for business processes with collaboration and wiki-like features embedded in [6], and an ad hoc workflow system focusing on non-intrusive capturing of human interactions in [5].

While most of the existing works in the social BPM area focus on adding social features to an existing BPM framework, our architectural framework [3], *Processbook*, gives principles and guidelines for managing personal processes *within* a social network. It embeds four key capabilities in its underlying social network: collaborative process modelling, knowledge capturing and sharing, social-network-based recommendation and notification-based management of the dynamic environment. This paper is an attempt to realise the conceptual framework introduced in [3] focusing on the process knowledge discovery and recommendation in personal process domain.

4   Process Aware Personal Process Management System

Processbook aims (i) to make personal process management as effortless as possible for individuals and (ii) to utilise user participation to produce meaningful social collective data. The first step is to engage people to manage their processes through *Processbook*. Given that *Processbook* users are ordinary people rather than trained BPM designers or knowledge workers, they posses little or no knowledge or prior experience in process modelling and management. Therefore one of the main issues *Processbook* deals with is to find a way to provide support for individuals in managing their tasks while simultaneously taking advantage of their social participation to enrich its support. In Sect. 4.1 we propose a simple modelling interface based on the idea of ToDoLists to facilitate the modelling experience for non-technical users. The second step is to expedite the transfer of knowledge about processes among users. For this purpose, in Sect. 4.2 we propose a method to capture users process knowledge. Then in Sect. 4.3 we show how to share the process knowledge in form of process recommendation.

4.1   Modelling Interface

Traditional business process modelling lacks the required flexibility and agility when it comes to unstructured or ad-hoc processes. To break the rigidity of traditional modelling methods and to simplify their syntax for novice users, we propose a simple modelling approach that resembles the natural planning model our brain follows. There are five steps that our minds go through to accomplish any task: defining purpose and principles, outcome visioning, brainstorming, organising, and identifying next actions[2]. Similarly our proposed planning approach consists of five steps:

- Defining a *goal*; A goal is any desired result that requires one or more action steps. It is described in natural language and is mandatory for each plan e.g. "Gain admission to a PhD degree in computer science at UNSW".
- Defining a set of *constraints*; Constraints are sets of parameters and criteria that further elaborate goals. They can be *global* to describe the general parameters e.g. "Application deadline for PhD admission is 1 Dec 2013" or *local* to reflect personal visions on the goal e.g. "field of study: BPM".
- Gathering all required *tasks*; Determining the set of required tasks is the first step towards the desired outcome. A task is a single unit of work in the boundaries of a particular goal. A goal is achieved when enough of the right tasks have been performed successfully and some outcomes have been created that closely enough match the initial vision.
- Elaborating tasks; A short-list of tasks are specified and elaborated by linking them to local constraints or by adding annotations to them.
- Identifying next task to do; The order in which users want to carry out their listed tasks is the final planning step and is repeated until all tasks are carried out or the desired outcome has been achieved.

Formally we define a simple modelling interface called *ToDoList* to realise the idea of natural planning.

Definition 1. A *ToDoList* is a quadruple (I, G, T, C), where

- I is a unique identifier for the ToDoList
- G is a statement of the *goal* in natural language
- T is a set of tasks to be done to achieve the goal
- C is a set of constraints on the tasks and the overall planning.

Each task gives a natural language statement of one activity to be completed in achieving the goal. A task is a pair (I, D), where I is a unique identifier for the task and D is a description which can be either a natural language description of an atomic task or a reference to another ToDoList.

Tasks are first class entities in our proposed ToDoList modelling approach. They are entered by individuals or given to them through a recommendation mechanism (Sect. 4.3). Figure 2 illustrates a *task state diagram* in *Processbook*. Each task, at any given time, could be in one of the "planning", "carrying out", "carried out" and "captured" states.

---

[2] Allen, D.: Getting things done. penguin books (2001).

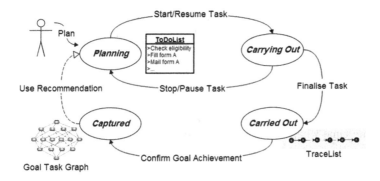

Fig. 2. Task State Diagram

- **Planning.** Once a task is defined or recommended to a user it will be put in the planning state. The planning state is similar to drafting a ToDoList and resembles the brainstorming step.
- **Carrying out.** Tasks are brought to "carrying out" mode on a user's decision to perform them. When a user identifies the next action she wants to take, she can simply pick the task from her ToDoList and bring it to the carrying out list. The task is then removed from the ToDoList, but can be reverted back when the user pauses or stops the execution of the task (to make changes to the task). When the user obtains the desired result from the running task, the task is considered finalised and will be moved to the "carried out" mode.
- **Carried out.** The carried out state consists of a set of completed tasks, ordered by their completion timestamp. Hence it could be regarded as a trace log for each ToDoList. It is expected that by the end of the personal planning, no tasks remain in the ToDoList or in the carrying out list. Instead the required tasks for an achieved goal are found in the carried out mode or more specifically in *TraceLists* that we will later define in Sec.4.2.
- **Captured.** If the user cancels her plan or does not achieve her specified goal, her traces will be deleted. Otherwise the trace is aggregated with other users' traces for the same goal. The aggregation produces a graph called *GTG Goal-Task Graph*, where users' successful experiences for a particular goal are captured. The *GTG* is described in 4.2, and captures the social production for a community of users who share the same goal.

4.2   Capturing Process Knowledge

Associated with each ToDoList, there exists a *TraceList* that is opened after the first task in ToDoList has been carried out and is closed after the corresponding ToDoList terminates. A TraceList captures the execution of tasks in a ToDoList.

Definition 2. A TraceList is a quadruple (I, G, CG, H) where

- I is a unique identifier for the TraceList

- G is a statement of the goal in natural language
- CG is a set of constraints on the goal
- H is a set of history records of tasks and their properties.

Each history record H consists of four elements:

- T is a unique identifier for the task
- CT is a set of constraints on the task
- ST is the time when task has been started
- ET is the time when task has been completed.

A ToDoList terminates successfully if all its tasks have been carried out and the owner of ToDoList confirms that a desired outcome has been reached by performing those tasks. The TraceList of such a ToDoList execution is called a *complete* TraceList. Intuitively a set of complete TraceLists for a certain goal will give a useful insight on how to achieve that particular goal. We argue that each TraceList resembles a blog post describing a personal solution to reach the goal, while the aggregation of TraceLists builds a wiki that describes the general solution to reach that goal in different contexts and from different perspectives. *Processbook* realizes the concept of social production by merging all complete TraceLists of a goal into a graph structure called GTG(Goal-Task Graph).

Definition 3. A GTG is a weighted directed graph $GTG(I, G, V, E, W, A)$ where

- I is a unique identifier for the GTG.
- G is the goal for which complete TraceLists are aggregated.
- V is the union of tasks in the complete TraceLists with goal G; the result set forms the vertices of GTG.
- E is the edges of GTG; each edge indicates the order of execution between two tasks.
- W is a weight associated with the edges of GTG indicating the number of times a particular edge has been followed over all TraceLists of goal G.
- C is a set of constraints associated with the edges of GTG indicating the circumstances under which a flow of tasks has occurred over all TraceLists of goal G.

For example, $T_i \xrightarrow{3,\{c_1 \vee c_2\}} T_j$ means task T_j preceded task T_i, three times in all complete TraceLists with either c_1 or c_2 mentioned as constraints of T_j in those complete TraceLists. Procedure 1 demonstrates the procedure of merging TraceLists into a GTG for goal G. The input of the procedure is a set of complete TraceLists that have G as their goal. `PrepareTraceList` orders tasks in each TraceList by their ET attribute. More complicated metrics could also be applied to include the ST and C attributes of history records as well, though it is out of the scope of this paper.

Each precedence relationship between two tasks is then added to the graph as follows: `FindTask`(T_i, T_j, GTG) searches the graph for tasks T_i and T_j.

If both tasks and their precedence relationship already exist in the *GTG*, we only need to increase the weight and update the constraint attributes of precedence. Otherwise if both tasks exist but they have not been connected, a new precedence should be added by (`AddPrecedence`) with its weight and constraints set. If only one of the tasks exists in *GTG*, we have to first add the missing task to the graph (`AddTask`), then add the corresponding precedence and finally set its weight and constraint attributes. This is similar to the case where both tasks are new to the *GTG*, with the minor difference that we have to add both tasks first.

4.3   Sharing Process Knowledge

Knowledge of achieving a goal, which we refer to as process knowledge, is captured when individuals carry out ToDoLists; this information is aggregated in the *GTG*. Process knowledge is the main artefact that is shared among users In *Processbook*. Process knowledge sharing happens via recommendations to targeted communities.

Figure 3 shows how such a sharing mechanism is realised in *Processbook*. For each goal, a community is created. Users are considered members of a community once they start towards a goal. These goal based communities, forming weak ties among users, are then used as a target group for recommendations. Users start planning their goals in the *ToDoList manager*. The executions of their plans are logged by *trace logger* in TraceList database. Secondary data from their involvements that includes votes and comments on recommended items are also recorded by the *event logger*. The *TraceList merger* performs aggregation of TraceLists (peer products) of a goal into the *GTG* to produce social production. Collective decisions are realised by applying users' votes on TraceLists and providing recommendations.

The *recommender* system consists of two different modules separated by the data sources they utilise: (i) *TraceList ranked retrieval module* that makes use of TraceList database and (ii) *process recommender* that utilises the *GTG*. Ranked retrieval of complete TraceLists could be based on several different metrics:

- Number of tasks in a TraceList; those with less tasks will be ranked higher implying they need less effort to achieve the goal
- Execution time of a TraceList calculated by $Max(ET) - Min(ST)$; those with shorter execution time will be ranked higher implying they reach the desired outcome sooner
- Popularity of a TraceList indicated by users'votes which reflects users' opinions on the usefulness and effectiveness of a TraceList.

The Process Recommender uses two slightly different approaches: recommending the *next best task* or recommending a *process path*. Both approaches use the *GTG* as their data source. The next best task recommender gives users a task at a time while being dynamically modified as the *GTG* grows. The Process Path Recommender, on the contrary, provides the whole set of tasks needed to reach the goal in the form of a path. It does not reflect *GTG* changes unless the user

Procedure 1. Merging *TraceLists* to GTG

Input: TRL: A set of complete *TraceLists* having goal G
Output: GTG for goal G

```
PrepareTraceList(TRL)
for all tracelists trl in TRL do
    for all T_i → T_j in trl do
        Let w_ij = trl_weight_{T_i→T_j}
        Let c_ij = trl_constraints_{T_i→T_j}
        found = FindTask(T_i, T_j, GTG)
        switch found
            case both T_i and T_j were found in GTG
                if T_i → T_j exist in GTG then
                    Weight_{T_i→T_j} += w_ij
                    Constraints_{T_i→T_j} = Constraints_{T_i→T_j} ∨ c_ij
                else
                    AddPrecedence(T_i → T_j)
                    Weight_{T_i→T_j} = w_ij
                    Constraints_{T_i→T_j} = c_ij
                end if
            case T_i was found in GTG
                AddTask(T_j, G)
                AddPrecedence(T_i → T_j)
                Weight_{T_i→T_j} = w_ij
                Constraints_{T_i→T_j} = c_ij
            case T_j was found in GTG
                AddTask(T_i, G)
                AddPrecedence(T_i → T_j)
                Weight_{T_i→T_j} = w_ij
                Constraints_{T_i→T_j} = c_ij
            case neither T_i nor T_j were found in GTG
                AddTask(T_i, G)
                AddTask(T_j, G)
                AddPrecedence(T_i → T_j)
                Weight_{T_i→T_j} = w_ij
                Constraints_{T_i→T_j} = c_ij
    end for
end for
```

explicitly asks for an updated recommendation. One advantage of the next best task recommender appears when a user wants to select a task manually, which is not necessarily the next best one. In this case the recommender will consider calculating the best remaining path, taking into account what user has chosen so far. In both cases, the user may fully or partially accept the recommendation or may completely reject it and continue making her own plans.

The major advantage of the process recommender over the TraceLists ranked retrieval module is that it better reflects the knowledge of the crowd. It is because

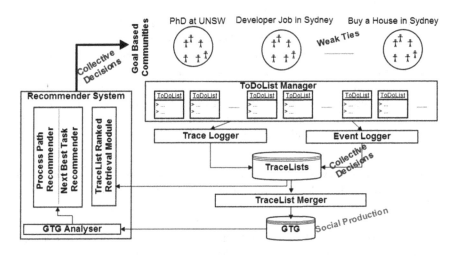

Fig. 3. Realisation of social software principles in *Processbook*

the next task or the process path given by the process recommender comes from merging different TraceLists and is considered to give the best possible solution from the crowd experience, while in the ranked retrieval system we are limited to the number of single TraceLists. The advantage of using a retrieval system based on single TraceLists is that what is finally recommended is already used by one or more users in real world, whereas in process recommender, the final recommended path might have not been carried out before by any user. In other words, it is a system generated path that is composed of different best practices found in the most successful attempts.

The basis of the process recommender is to find the minimum weighted path in the *GTG*. Weights in the *GTG* represent the popularity of a task flow but they have to be normalised so that Dijkstra's algorithm could be applied and the minimum weighted path from the starting task to the end task found. We also elaborate the weights of the *GTG* by taking into account users' votes in addition to the popularity measure. Therefore the final weight for an edge in *GTG* between tasks T_i and T_j is calculated as follows:

$$\alpha \times (10 - \frac{w_{ij}}{N} \times 10) + \beta \times (10 - vote)$$

where w_{ij} is the initial popularity weight of the edge between T_i and T_j, N is the number of complete TraceLists merged into the *GTG*, *vote* indicates the average votes for the task flow, ranging from 0 to 10 with 10 meaning the best and α and β are coefficients for popularity metrics and users' votes respectively.

It is probable that tasks that have been repeatedly used in complete TraceLists be excluded from the final recommendation due to existence of some unrealistic short paths in the graph. To avoid ignoring such tasks and thus improving the recommendation results, we define the concepts of *required tasks* and *required ratio*. The required ratio for task T_i is calculated by w_{*i}/N where

w_{*i} is the sum of the weights of ingoing edges to task T_i and N is the number of complete TraceLists merged into the GTG. A threshold for required ratio is set to force the inclusion of task T_i. Tasks whose required ratio is above the threshold are called required tasks, implying they are the necessary steps if the process is to achieve the goal. Considering required tasks, the recommendation mechanism should also be modified. To this purpose, the GTG *analyser* shown in Fig. 3 is responsible for marking required tasks in each GTG. The final recommended paths will be filtered to avoid ignoring required tasks.

Towards Context-Aware Process Recommendation. A further step in sharing the process knowledge is to enable context-aware process recommendation by taking into account user profiles and task constraints. The pre-requisite is to have a logical mapping between user profile attributes and task constraints. For instance, in a personal process which is about applying for a PhD scholarship, users should fill out the required fields about their educational background such as degrees they hold, universities and schools they have attended, etc.

Having a set of constraints associated with each edge in GTG in Procedure 1, first challenge would be to find the most relevant constraint(s) for each edge. A simple approach is to rank those constraints based on the frequency of their appearance in the trace logs. This way, in its simplest form, we would have a GTG with each of its edges labelled with the most frequent constraint. Subsequently any of the recommendation algorithms introduced in Sec.4.3 can be modified to include constraint labels in their internal logic.

An alternative approach is to first cluster trace logs based on the user profiles and then run Procedure 1 to build GTGs for each cluster. An agglomerative hierarchical clustering algorithm can be used to find the proper number of clusters. Prior to using any recommendation algorithm, we should first assign each user a cluster of which its center is nearest to her user profile. The input of the recommendation algorithm would be the GTG which is built over that particular cluster.

5   Implementation and Pilot Study

A prototype of the proposed system is implemented as a Java-based Web application. In this section, we demonstrate our solution by presenting a simplified version of test cases we have run with *Processbook*. We have asked five students in our school to plan for the case "Applying for UNSW PhD scholarship" through ToDoList modelling interface of *Processbook*. Based on this input we show how the system generates recommendations for Ali, who also wants to apply for a PhD scholarship.

Users' knowledge and experience in planning the scholarship case has been captured in TraceList database. Task descriptions, complete TraceLists and votes statistics are provided in Fig. 4. *Processbook* aggregates TraceLists and build the GTG shown in Fig. 4. Without loss of generality, we assume that user votes for each edge is equal (e.g. 5 out of 10). Moreover since the effect of constraints

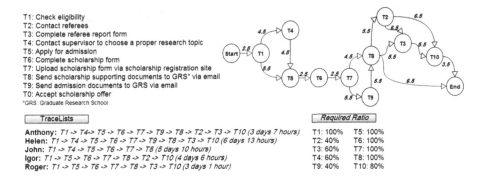

T1: Check eligibility
T2: Contact referees
T3: Complete referee report form
T4: Contact supervisor to choose a proper research topic
T5: Apply for admission
T6: Complete scholarship form
T7: Upload scholarship form via scholarship registration site
T8: Send scholarship supporting documents to GRS* via email
T9: Send admission documents to GRS via email
T0: Accept scholarship offer
*GRS: Graduate Research School

TraceLists		Required Ratio	
Anthony: $T1 \to T4 \to T5 \to T6 \to T7 \to T9 \to T8 \to T2 \to T3 \to T10$ (3 days 7 hours)		T1: 100%	T5: 100%
Helen: $T1 \to T4 \to T5 \to T6 \to T7 \to T9 \to T8 \to T3 \to T10$ (6 days 13 hours)		T2: 40%	T6: 100%
John: $T1 \to T4 \to T5 \to T6 \to T7 \to T8$ (5 days 10 hours)		T3: 60%	T7: 100%
Igor: $T1 \to T5 \to T6 \to T7 \to T8 \to T2 \to T10$ (4 days 6 hours)		T4: 60%	T8: 100%
Roger: $T1 \to T5 \to T6 \to T7 \to T8 \to T3 \to T10$ (3 days 1 hour)		T9: 40%	T10: 80%

Fig. 4. "Applying for UNSW PhD scholarship" TraceLists and GTG

has not yet implemented in the recommender system of *Processbook*, we ignore constrains labels of GTG edges.

In terms of number of tasks, John's TraceList stands higher while in terms of execution time, Anthony's TraceList would be recommended to Ali. However, if Ali decides to use a process path recommender, he will be offered a new path which does not exist in any of those TraceLists. The recommended path excludes T_9 which as we detected is a common misunderstanding between applicants. However the path shown in Fig. 4 also excludes T_3, T_4 and T_{10}, all seems mandatory tasks according to UNSW policies. To avoid this undesirable elimination, we have to tune the required threshold to 50 %, to enforce inclusion of such required tasks. Therefore *Processbook* filters out the path illustrated in Fig. 4 and returns the optimal path as can be seen in an screenshot of the system in Fig. 5.

5.1   Preliminary Evaluation of Process Recommendation Algorithms

A main part of our proposed personal process management system is the recommendation module which plays a significant role in process knowledge sharing. A preliminary experiment was conducted to measure the effectiveness of the core recommendation algorithms introduced in Sec.4.3. For a proper evaluation we needed a fairly large amount of traces which was impossible to acquire from prototype logs in the limited time we had. Moreover, to the best of our knowledge there is no dataset available online to test process recommendation systems (or any path-based recommendation systems). As a result, we developed a TraceList simulator to produce TraceLists based on user profiles and expertise. Users' expertise parameter was added to enable generation of traces of different qualities: from nearly perfect traces to complete random task selection.

Two different algorithms were implemented for the purpose of this experiment: (i) GTG: created by Procedure 1 labelled with the most frequent constraint, and (ii) clustered GTG: in which trace logs were clustered based on user profiles and Procedure 1 was executed for each cluster afterwards. We used

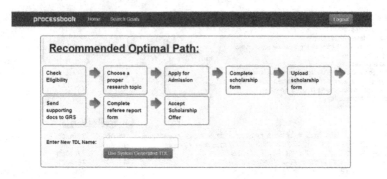

Fig. 5. Screenshot of optimal path generated by *Processbook*

TraceList simulator to produces 50, 200, 500 and 1000 successful traces for Scholarship application process. Accuracy of each recommendation algorithm was
measured by calculating the similarity between the recommended path and the
ground truth (i.e. the optimised path to apply for scholarships at UNSW). The
similarity is formally calculated as follows:

$$1 - \frac{editDistance(r, g)}{max(length(r), length(g))}$$

where r and g are recommended path and ground truth respectively and length
indicates the number of tasks in a path. Table 1 summarizes the accuracy results
obtained for both GTG and clustered GTG algorithms.

Table 1. Accuracy of recommendation algorithms

Size of log	GTG	clustered GTG
50	0.715	0.823
200	0.718	0.914
500	0.723	0.920
1000	0.731	0.917

The experiment shows that process paths recommended by clustered GTG is
significantly more accurate than those recommended by GTG. However, we wish
to emphasise that the experiment reported above is at an early stage and can
be improved by (i) acquiring the real world traces and (ii) adding social network
aspects such as users' votes and communities. It should also be extended to test
variations of recommendation algorithms against various personal processes that
differs in terms of size and complexity.

6 Conclusion and Future Work

Our proposed solution for personal process management is to create a flexible process management environment within a social network structure. We maintain the process awareness of traditional BPM to be able to handle dependencies and constraints between tasks, but at the same time we follow the principles of social software by establishing goal-based communities, enabling social production and utilising collective decisions. To realise all these, we have proposed a novel method for modelling personal processes based on the idea of ToDoLists and have implemented a mechanism to unobtrusively capture users' experience separately and then aggregate them in a graph structure that can be used as a source for process recommendation.

Our future work includes: (i) improving context-aware process recommendation, (ii) enforcing trustworthy recommendation by utilising more social software features and (iii) enriching our capturing mechanism by enabling parallelism in task flows. As well, we intend to evaluate *Processbook* in real world scenarios by conducting comprehensive user studies. We believe that *Processbook* can be employed in many knowledge intensive domains in addition to personal process management.

References

1. Brambilla, M.: Application and simplification of BPM techniques for personal process management. In: La Rosa, M., Soffer, P. (eds.) BPM Workshops 2012. LNBIP, vol. 132, pp. 227–233. Springer, Heidelberg (2013)
2. Bruno, G., Dengler, F., Jennings, B., Khalaf, R., Nurcan, S., Prilla, M., Sarini, M., Schmidt, R., Silva, R.: Key challenges for enabling agile bpm with social software. J. Softw. Maint. **23**(4), 297–326 (2011)
3. Hajimirsadeghi, S.A., Paik, H.-Y., Shepherd, J.: Processbook: towards social network-based personal process management. In: La Rosa, M., Soffer, P. (eds.) BPM Workshops 2012. LNBIP, vol. 132, pp. 268–279. Springer, Heidelberg (2013)
4. Koschmider, A., Song, M., Reijers, H.A.: Social software for modeling business processes. In: Ardagna, D., Mecella, M., Yang, J. (eds.) Business Process Management Workshops. LNBIP, vol. 17, pp. 666–677. Springer, Heidelberg (2009)
5. Martinho, D., Silva, A.R.: Non-intrusive capture of business processes using social software - capturing the end users' tacit knowledge. In: Daniel, F., Barkaoui, K., Dustdar, S. (eds.) BPM Workshops 2011, Part I. LNBIP, vol. 99, pp. 207–218. Springer, Heidelberg (2012)
6. Silva, A.R., Meziani, R., Magalhães, R., Martinho, D., Aguiar, A., Flores, N.: AGILIPO: embedding social software features into business process tools. In: Rinderle-Ma, S., Sadiq, S., Leymann, F. (eds.) BPM 2009. LNBIP, vol. 43, pp. 219–230. Springer, Heidelberg (2010)
7. Weber, I., Boualem Benatallah, H.P.: Form-based web service composition for domain experts. ACM Trans. Web 8(1), Article 2, pp. 1–40 (2013)

Discovering Information Diffusion Processes Based on Hidden Markov Models for Social Network Services

Berny Carrera, JinSung Lee, and Jae-Yoon Jung[✉]

Department of Industrial and Management Systems Engineering,
Kyung Hee University, 1 Seochen-dong, Giheung-gu, Yongin, Gyeonggi,
Republic of Korea
{berny,jinsll27,jyjung}@khu.ac.kr

Abstract. As social network services (SNS) such as Facebook and Twitter have become a major means of communication on the Internet, many studies have been conducted using SNS data. Yet, the studies still have limitations in the viewpoint of dynamics of information. For example, although a variety of social network analytics visualize the relationship among people, they mainly focus on the static relationship rather than dynamic information flow on the social network. In this research, we introduce the Hidden Markov Model for Information Diffusion (HMMID) that applies HMM to process mining techniques in order to discover and visualize information flow from SNS log data. The proposed methodology helps to visualize the sequences and paths of information delivery among users from reality of SNS event log along with their probabilities on the arcs. Experiments with synthetic data and real-life Facebook data were conducted using the HMMID Miner, which have been developed on the ProM framework. The methodology is illustrated with the two kinds of experiments to show how to get knowledge on the information dynamics inside the social network.

Keywords: Social network analytics · Process mining · Probabilistic process discovery · Hidden markov models · Information diffusion

1 Introduction

Nowadays, social network service (SNS) is a representative service of the Web 2.0 and it is widely used in the world. Online social network services such as Facebook and Twitter have hundreds of millions of users, and each user in these services are connected to their friends, family, and co-workers. Many people commonly communicate their opinion or news through SNS on the Internet [1]. As SNS has become a major means of online communication, various studies on online social network analysis have been progressing [2–4].

However, studies on online social networks are still insufficient to understand information flow in SNS and predict the behaviors of users. Considering the aspect of visualization, we can use the social network that shows the relationship of people in society in the area of social science to represent the users' behaviors in social media.

© Springer International Publishing Switzerland 2015
J. Bae et al. (Eds.): AP-BPM 2015, LNBIP 219, pp. 170–182, 2015.
DOI: 10.1007/978-3-319-19509-4_13

One can understand the structure of social relationship such as distinction of community, relationship pattern, and tracing diffusion through analyzing the diagram [5]. Although such social network analytics provide good methods for understanding information diffusion on the SNS, they cannot visualize effectively the sequence and path of the information propagation inside a social network.

In this research, we propose a process discovery methodology based on Hidden Markov Model (HMM) for easy understanding of information diffusion and by discovering a process map from SNS data (i.e. social event logs). In general, HMM is a methodology that stochastically estimates the parameters of hidden states from observations, and it is used widely in speech recognition, data recognition and bioinformatics [6]. Meanwhile, process mining techniques extract knowledge from event logs in enterprise information systems. These techniques mainly aim at process discovery, conformance checking and operational support in order to improve a business processes [7].

The proposed technique, named the Hidden Markov Model for Information Diffusion (HMMID), has the following advantages compared to traditional social network analysis tools. First, the methodology derived process maps from the event log which contains information such as event time, activity, and resource. The information can be used to represent information delivery and acquisition on the network, and the direction of information delivery can be represented in paths to a specific user group. Second, just as HMM, all the possible states in which information is transmittable can be analyzed based on the discovered probabilistic process model.

In the remainder of the paper, we first introduce the related work on information diffusion of social networks using process mining techniques in Sect. 2. In Sects. 3 and 4, we explain the concept and the algorithm of HMMID, and the calculation of HMMID parameters. In Sect. 5, two experiments with a synthetic data and a real-life Facebook log are conducted to illustrate the proposed methodology. Both experiments are analyzed by using the HMMID miner which has been developed on the ProM framework. Finally, we draw conclusions in Sect. 6.

2 Related Work

Recently, some studies were conducted to analyze a social network by using process mining techniques. In van der Aalst et al. [8], the relation data among workers from event log were analyzed and presented in the form of social network diagrams. Their approach is not suitable for analyzing online social network data since it is focused on analyzing the organizations under business process management in an enterprise.

Some other studies were conducted to build an information diffusion model using real-life social network data. Kim et al. [9] presented an information diffusion model by using data of a social blog service Tumblr and analyzed the structure of the model. Also, further research to discover information flow in Facebook sites was conducted [10]. In Obregon et al. [11], a hierarchical process discovery method was proposed to understand the information diffusion process among user groups. Compared to a highly complicated process model from social network data, these results enabled people to understand the information flow better by presenting the abstract process model.

However, those information diffusion models can be used to infer the information diffusion because they did not consider the probabilities of the state transition in social networks. The proposed methodology in this research, HMMID, represents all possible states of information delivery between SNS users based on SNS data, and it enables to conduct probabilistic analysis on the sequences and paths, as well.

3 Hidden Markov-Based Model for Information Diffusion

In this section, the formal models of the proposed method are first described in Sect. 3.1. The principal challenge is to extract proper log data from SNS in order to discover effective process models of representing information diffusion flow. The mining algorithm for HMMID, which is an extension of typical HMM to social network analytics, is then presented in Sect. 3.2. From the SNS log data, some information among actors are investigated to analyze their relations and the paths on the network.

3.1 Social Network Service Data

A SNS event log, which will be used an input for our proposed algorithm, is first described to understand the principal characteristics of information diffusion in SNS. It is assumed that actors will react the posts written in the same SNS site in similar ways, and under the assumption the flow of actors' actions for the posts will be discovered as an process model for information diffusion for the target SNS site. For example, a post in a Facebook fan page contains ordered actions of users, such as comments, likes, and shares. Hence, a SNS log simply consists in multiple sequences of actors who have reacted for the posts in the target site. In each post, one can observe a sequence of actors that have reacted with comments, and the sequence of actors is called an action trace in this research. Herein, the actors' actions in the trace are ordered by their timestamps such as time of like, time of comment or time of share.

Definition 1. (*SNS log*). For a given social network service (SNS), let $V = \{v_1, \ldots, v_M\}$ be a finite set of observed actors and $P = \{\rho_1, \ldots, \rho_K\}$ be a finite set of posts written in the SNS. An action trace of actors for a specific post ρ_k, denoted by σ_k, is a sequence of actors in $v \in V$ i.e. $\sigma_k = <v_1, \ldots, v_{n(k)}>$ for $1 \leq k \leq K$. A SNS log, denoted by $L = [\sigma_k]$, is a multi-set of action traces over V and P in the SNS.

A SNS log is used to reveal the ordered relationship among individual users or the circles of friends. Based on the reality stored in the log, we will construct process models that can interpret or explain the behaviors in terms of information diffusion.

3.2 HMMID Miner

The goal of HMMID is to analyze the sequence of actors' actions to find the hidden states of interaction using HMM [12]. Actually, we cannot know who have made a specific user read the original post, but we can observe only the sequence of acted users according to the time order. In the viewpoint of HMMID, the hidden states represent the information delivery from one user to another, whereas the observations represent their traces in many posts.

A SNS log includes a variety of sequences of interactions between actors which can be used to mine the actual relationships during users' interaction [13]. An information diffusion state represents the delivery or the spread of information from one actor to another. From a SNS log, it can be extracted as follows:

Definition 2. (*Information diffusion state*). Let V be actors in a SNS log L. An information diffusion state over L, denote by $s = v_i \rightarrow v_j$, is an ordered pair of adjacent two actors in V that appear in any action trace $\sigma \in L$. Furthermore, a set of information diffusion states is denoted by S.

In order to discover an information diffusion model, an interaction matrix is constructed to evaluate the frequencies of each information diffusion states from a given SNS log. In a word, the frequency of each state can be calculated from every observed sequences in posts of the log. The interaction matrix will be used to extract the probabilities of state transitions in the information diffusion model.

Definition 3. (*Interaction matrix*). Let V be actors and S be information diffusion states of a SNS log L. An interaction matrix is $T = (\tau_{ij})$ where $\tau_{ij} = \sum_{\sigma \in L} |v_i \rightarrow v_j|$ is the sum of the frequencies of an information diffusion state $s = v_i \rightarrow v_j \in S$ in every action trace $\sigma \in L$.

In our approach, an interaction matrix is required to construct the state transitions of the HMMID by analyzing the relations of response and the friendship of a pair of actors. The matrix is focused on a process perspective, especially in the analysis and differentiation of actors and their relationship. It can measure the frequency of actors' appearances (i.e., when *actor(b)* replies to *actor(a)* inside the sequences in different posts. This model is suitable for real situations (e.g., SNS log) and it can be applied in filtering the data with a threshold to reduce weak links between actors and presenting a more accurate and understandable model.

The next step after the construction of the interaction matrix is to start with the construction of the HMMID that models an information diffusion graph based on the states and observations of the HMM. In the HMMID diagram, circle-shaped states represent the information delivery between actors, and transitions are represented by two different kinds of arcs: one for the transition between states and the other for the actual observation of an actor for a state.

The states represents the spread of information diffusion and in terms of social network, the relationship of communication between the actors. For example, the information diffusion of '*actor(b)* $\rightarrow$ *actor(c)*' has a causality relation with '*actor (c)* $\rightarrow$ *actor(b)*' and '*actor(c)* $\rightarrow$ *actor(d)*' and therefore creates a transition denoted by an "arc" from '*actor(b)* $\rightarrow$ *actor(c)*' to these information diffusion states. The observations are the actors related in the intercommunicated network.

In order to indicate the beginning and ending of the spread of information diffusion, the HMMID added the *start* and *end* dummy states. The *start* state is the initial state that a user wrote the original post, and the initial state probability distribution π of at the initial state is straightforward, i.e. $\pi = (1, 0, \ldots, 0)$. An observation (i.e., actor) is represented by a square shape with the actor's name. To model the termination of the conversation, the *end* state is added which receives its input from the last information diffusion state, in each sequence and from which no other state is reachable. The formal model of HMMID can be described as follows:

Definition 4. (*HMMID*). A HMMID for a SNS log L, which is an extension of HMM of representing information diffusion, is $\lambda(L) = (S', V', A, B, \pi)$ where S' is a set of states, V' a set of actors, A the matrix of state transition probability distribution, B the matrix of observation symbol probability distribution, and π the transition probability distribution for the initial state. Note that $\sum_j a_{ij} = 1$, $\sum_{v_m} b_i(v_m) = 1$ and $\sum_i \pi_i = 1$.

$S' = \{s_i\} = S \cup \{s_{start}, s_{end}\}$ for $1 \le i \le N + 2$, where S is the information diffusion state set of L, and s_{start} and s_{end} are two dummy states.

$V' = \{v_m\} = V \cup \{end\}$ for $1 \le m \le M + 1$, where V is the actor set of L and *end* is a dummy observation.

$A = (a_{ij})$ for $1 \le i, j \le N + 2$, where a_{ij} is the probability of state transition from s_i to s_j

$B = (b_i(v_m))$ for $1 \le i \le N + 2$, $1 \le m \le M + 1$, where $b_i(v_m)$ is the probability of observing v_m in state b_i.

$\pi = (\pi_i) = (1, 0, \ldots, 0)$ for $1 \le i \le N + 2$, where π_i is the probability of being in s_i at time 1, i.e. $\pi_1 = \pi(s_{start}) = 1$ and $\pi_i = 0$ for all other i's.

As described in Fig. 1, the HMMID learns from an SNS log which is fully observed and includes the sequence of actors and their replies in different posts. This information is analyzed and modeled in a HMM. The states of the HMM can be represented based on the information diffusion of the comments. The HMMID algorithm analyzes and

---

```
1:   Input: SNS log L = [σ].
2:   Initialize
3:   A set of actors V={}. A set of information diffusion states S={}.
4:   For each action trace σ in L
5:       Insert all actors in σ to V.
6:       Insert all the ordered pairs of σ into S.
7:   End For
8:   Create S' = S∪{s_start, s_end} and V' = V∪{end}.
9:   Create an empty |S'|×|S'| interaction matrix T.
10:  For each state s in S'
11:      Increase τ_ij by 1 for every ordered pair s = v_i → v_j in σ.
12:  End For
13:  Create a |S'|×|S'| matrix A= (a_ij) by calculating the state transition
     probability a_ij for every pair of states s_i and s_j.
14:  Create a |S'|×|V'| matrix B= (b_i(v_m)) by calculating the observation
     symbol probability of v_m in state s_i.
15:  Create a |S'| vector π =(1, 0, ..., 0)
16:  Output: initial HMMID: λ = (S', V', A, B, π )
```

Fig. 1. HMMID miner algorithm.

models the information flow among actors such as their friends and neighbors, and it considers the reply order among different comments.

In the next section it is described how to estimate the parameters (i.e. probabilities) of the HMMID. Two methods are introduced through which the optimal parameters can be induced based on the log data.

4 Parameter Calculation

Two ways of estimating the parameters of HMMID from SNS log, maximum likelihood (ML) and expectation maximization (EM) [12, 14], are introduced in this section. ML can be adopted for parameter estimation to infer exactly the information diffusion states with their respective observations. The parameters are calculated from the mined information diffusion states including the actors and the frequency of occurrence in order to calculate the interaction matrix. Contrarily, what happens if it is not possible to calculate initial values for the state transition probability distribution? In this case, EM can be used to estimate the parameters so that the initial variables are calculated randomly for two probability distribution matrices of the model, A and B. After a few training runs with E and M steps, the expectation maximization for these probabilities are given in two optimal matrices, $A*$ and $B*$. The parameter estimation of HMM has been discussed for resource allocation problems in our previous study [15], while the parameter estimation is explained for information diffusion in this paper.

4.1 Maximum Likelihood

To calculate the parameters, it is necessary to analyze the HMMID including the elements such as S', V', and π from the SNS log. The next step is to determine the initial state transition probability distribution A and the observation symbol probability distribution B. These steps were calculated with the HMMID miner algorithm. To analyze both probability distributions, the frequency f of each hidden state (information diffusion state) and observations (actors) are mined and the weight of the arcs are calculated. Starting with the transition probabilities, the interaction matrix is needed to obtain the frequencies of the information diffusion states.

For a given $\lambda(L) = (S', V', A, B, \pi)$, $\lambda'(L)$ is the set of all possible ordered pairs of an actor and a state over $\lambda(L)$, i.e. $\lambda'(L) = \{(v_1, s_1), \ldots, (v_Q, s_Q)\}$ for $v_q \in V'$ and $s_q \in S'$ in L. Such pairs of observations (i.e. actors) and hidden states (i.e. information delivery) are represented by a HMM. The probability of λ', denoted by $P(\lambda')$, is calculated:

$$P(\lambda') = \prod_{q=1}^{Q} P(v_q, s_q). \tag{1}$$

$$P(v_q, s_q) = \prod_{i} \pi_i^{f(i, v_q, s_q)} \prod_{i,j} d_{ij}^{f(i,j, v_q, s_q)} \prod_{i, v_m} b_i(v_m)^{f(i, v_m, v_q, s_q)}. \tag{2}$$

In Eq. (2), $f(i, v_q, s_q)$ is the frequency of state s_i that follows the $\{start\}$ state in (v_q, s_q), $f(i, j, v_q, s_q)$ is the frequency of state s_i that precedes state s_j, which can be calculated in the interaction matrix, and $f(i, v_m, v_q, s_q)$ is the frequency of state s_i that is paired with actor v_m. Note that a_{ij} and $b_i(v_m)$ are the probabilities in the matrices A and B of λ', respectively, and π_i is the probability of state s_i in the $\{start\}$ state in L.

The logarithm of probability of λ', denoted by $\mathrm{Log}(\lambda')$, can be easily calculated:

$$\mathrm{Log}(\lambda') = \mathrm{Log}(P(\lambda')) = \log \prod_{q=1}^{Q} P(v_q, s_q)$$
$$= \sum_{q=1}^{Q} \log P(v_q, s_q) \tag{3}$$

$$= \sum_{q=1}^{Q} \sum_{i} f(i, v_q, s_q) \log(\pi_i) + \sum_{i,j} f(i, j, v_q, s_q) \log(a_{i,j})$$
$$+ \sum_{i, v_m} f(i, v_m, v_q, s_q) \log(b_i(v_m)).$$

In order to find λ'^* of maximizing $\mathrm{Log}(\lambda')$, the following formula is needed: $d\,\mathrm{Log}(\lambda')/d\lambda' = 0$.

$$\pi_i^* = \frac{\sum_q f(i, v_q, s_q)}{\sum_q \sum_h f(h, v_q, s_q)}. \tag{4}$$

$$a_{ij}^* = \frac{\sum_q f(i, j, v_q, s_q)}{\sum_q \sum_h f(i, h, v_q, s_q)}. \tag{5}$$

$$b_i(v_m)^* = \frac{\sum_q f(i, v_m, v_q, s_q)}{\sum_q \sum_{v_m' \in V'} f(i, v_m', v_q, s_q)} \tag{6}$$

The three parameters can be calculated with a supervised learning from the mined information diffusion states and actors, allowing for the calculation of the probabilities of the states and observations. Equations (4), (5) and (6) explain the calculation of parameters in ML. π_i, the frequency of state s_i, is divided by the total frequency of all states that precede $\{start\}$ state, $a_{i,j}$, the frequency of state s_j, follows state s_i, is divided by the total frequency of all states that follows state s_i, and $b_i(v_m)$, the frequency of state s_i that is paired with an actor v_m, is divided by the total frequency of all actors paired with state s_i.

4.2   Expectation Maximization

Expectation Maximization (EM) maximizes the probabilities of observed actors in HMMID. Contrary to ML, EM evaluates the current parameters in the E step and recalculates the parameters to maximize the probabilities in the M step. EM maximizes the parameters of the given model repeatedly using an input data to find the maximum likelihood of the parameters. EM can be used in two cases: if we have the previous HMMID $\lambda = (S', V', A, B, \pi)$, we can recalculate the parameters with an actual SNS log through EM. But, if we have just the sequences of the actors from a SNS log, we can attempt to find the new parameters of maximizing the probabilities of the observed sequences. It have to be started from a random values for $\lambda = (S', V', A, B, \pi)$.

As described in Fig. 2, the EM procedure essentially calculates the parameters for A, B and π in two steps, E and M. These parameters are calculated using forward and backward algorithms from HMM [16].

1:	**Inputs:** $\lambda = (S', V', A, B, \pi)$ and SNS log L
2:	**Repeat**
3:	Using the forward algorithm, calculate α_i for each trace σ in L
4:	Using the backward algorithm, calculate β_i for each trace σ in L
5:	(Re-)Estimation of temporal variables, γ and ξ
6:	Calculate γ_i and ξ_{ij} based on $\sum \alpha_i$ and $\sum \beta_i$
7:	Calculate A^*, B^*, and π^* from γ and ξ
8:	Updating λ
9:	**Until** A and B do not change
10:	**Output:** $\lambda^* = (S', V', A^*, B^*, \pi^*)$

Fig. 2. Expectation maximization procedure for HMMID.

To obtain $\lambda^* (= S', V', A^*, B^*, \pi^*)$, the EM procedure uses two variables, α_i and β_i. α_i computes the probability given a sequence of actors in λ for state s_i in time t and it is (re-)estimated from the forward algorithm. β_i is the probability for state s_i at time $t + 1$ given a sequence of actors and it is (re-)estimated from the backward algorithm.

The procedure also utilizes two temporal variables, γ_i and ξ_{ij}, [16] to estimate the values of A^*, B^*, and π^*.

$$\gamma_i(t) = \frac{\alpha_i(t)\beta_i(t)}{\sum_{q=1}^{Q} \alpha_q(t)\beta_q(t)}. \tag{7}$$

$$\xi_{ij}(t) = \frac{\alpha_i(t)a_{ij}\beta_j(t+1)b_j(v_{t+1})}{\sum_{p=1}^{n}\sum_{q=1}^{n} \alpha_p(t)a_{pq}\beta_q(t+1)b_q(v_{t+1})}. \tag{8}$$

These temporal variables are useful to estimate the maximum likelihood of the parameters for $\lambda(L)$. $\gamma_i(t)$ can be interpreted like the possibility to be in a specific information diffusion state on a determined time of the process, and $\xi_{ij}(t)$ the probability of the following information diffusion states in a specified time. So, for a given observation sequence of comments in a post, $\gamma_i(t)$ is the probability of being in the information diffusion state s_i at time t, and $\xi_{ij}(t)$ is the probability of being in the information diffusion state s_i at time t and being in the information diffusion state s_j at time $t + 1$.

$$a_{ij}^* = \frac{\sum_{t=1}^{T-1} \xi_{ij}(t)}{\sum_{t=1}^{T-1} \gamma_i(t)}. \tag{9}$$

a_{ij}^* is the expected probability of being in the information diffusion state s_i to s_j. a_{ij}^* is the expected number of transitions for the information diffusion states and the probability distribution of the frequency that the information diffusion state s_i occurs in sequence from $t = 1$ to $T - 1$.

$$b_j^*(v_m) = \frac{\sum_{t=1;v_t=v_m}^{T-1} \gamma_j(t)}{\sum_{t=1}^{T-1} \gamma_j(t)}. \tag{10}$$

$b_j^*(v_k)$ is the predicted parameters of the actors assigned to a specified information diffusion state. $b_j^*(v_m)$ is the expected observation symbol probability distribution, therefore it is the expected number of times in the information diffusion state s_j and observing the actor in the post v_m, divided by the expected number of times in s_j.

$$\pi_i^* = \gamma_1. \tag{11}$$

π_i^* is the expected initial state distribution at time 1 in state i. π_i^*, is the expected probability for all states s_i that precede $\{start\}$ state.

5 Implementation and Experiments

The proposed method was implemented as a plug-in on the ProM framework, called HMMID Miner. The plug-in takes SNS log as an input data and constructs an information diffusion HMM from the SNS log using the algorithm defined previously with the calculations of their parameters.

We demonstrate the applicability of our approach with two different experiments to describe how our method can be applied to social networks. Section 5.1 describes the experiment with synthetic data and Sect. 5.2 shows the method's applicability to real-life SNS log. The latter experiment was performed by using Facebook data from one of the leaders in news and information around world, the CNN fan page.

5.1 Experiment 1 – Synthetic SNS Data

A simple synthetic data used for the first experiment is shown in Table 1. The data contains 6 posts and 27 comments which were written by five actors $\{a, b, ..., e\}$. A text of a comment is described in the subscript of its actor in the data, e.g. $a_{\text{incididunt ut labore}}$ has a text of 'incididunt ut labore' written by actor a. A post is a sequence of actors, e.g. σ_{546281} has a sequence of four actors, $< a, b, c, d >$.

As shown in Fig. 3, a HMMID was constructed from the synthetic SNS log in Table 1. The HMMID includes two types of nodes: information diffusion states depicted by circles and actors depicted by rectangles. In the information diffusion states

Table 1. Example of SNS log.

Post ID	Comments
546281	aLorem ipsum dolor... , $b$sit amet,... , $c$consectetur adipiscing elit, , $d$sed do eiusmod tempor
546282	$a$incididunt ut labore, $b$et dolore magna aliqua. , cUt enim ad minim, $d$veniam, quis nostrud exercitation
546283	$a$ullamco laboris, $b$nisi ut aliquip ex, $c$ea commodo, $b$consequat, cDuis aute, $d$irure dolor
546284	$a$in reprehenderit, $c$in voluptate, $b$velit esse, $c$cillum dolore eu, $b$fugiat nulla pariatur, dExcepteur sint
546285	$a$occaecat cupidatat non, $c$proident, sunt in, $b$culpa qui officia deserunt, $d$mollit anim id
546286	$a$est laborum, eAt vero eos et accusamus, $d$et iusto odio dignissimos ducimus

the actor's actions such as user comments are analyzed, and the edges of the model serve as a direction for the information propagation with an estimated probability. The information diffusion states are related to different actors obtained from the SNS log. The dotted lines represent the probability and relation with the states.

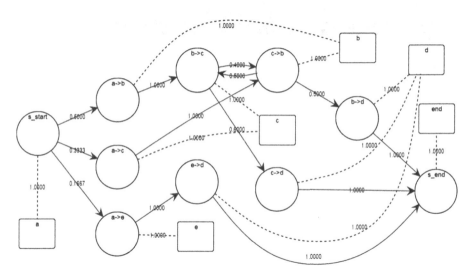

Fig. 3. HMMID constructed from the synthetic dataset.

With the constructed HMMID we can observe that commonalities exist between state '$b \rightarrow c$' and '$c \rightarrow b$', and we can measure the probability of communication from *actor(b)* and *actor(c)*.

There are two questions that can be answered based on the discovered HMMID. The first question is: what is the probability of a given sequence of actors? In the result of synthetic data, the question can be analyzed by using the forward algorithm, for example, the probability of the sequence $<a \rightarrow b \rightarrow c \rightarrow d \rightarrow End>$ is calculated with $P(a \rightarrow b \rightarrow c \rightarrow d \rightarrow End|\lambda) = 0.6 \times 1.0 \times 0.5 \times 1.0 = 0.3$. The second question is: given a sequence of actors, what is the most probable sequence of actors' actions in the information diffusion process? The question can be answered by using the Viterbi algorithm [12]. But, in this example, the sequence can be found intuitively by using the forward algorithm, instead of the Viterbi algorithm. The most probable sequence in the information diffusion process is <'*Start*', '*a* → *b*', '*b* → *c*', '*c* → *d*', '*End*'>.

5.2   Experiment 2 – Real-Life SNS Data

The methodology is next evaluated by using the Facebook data gathered from the CNN Facebook fan page. The data of the CNN fan page for one month, from May 1st to June 1st, 2014, was collected from all posts. Totally, 515 posts with 35 actors were associated with 2,705 comments. In order to reduce the noise (i.e. infrequent actors) the data was filtered to show the actors that appeared more than 100 times. This filtering represents around 20 % of the top post users. Also, all the actors that appear more than 2 times sequentially were described only one time. Finally, the posts which have more than 3 comments were filtered. A HMMID constructed from the CNN Facebook fan page data is shown in Fig. 4.

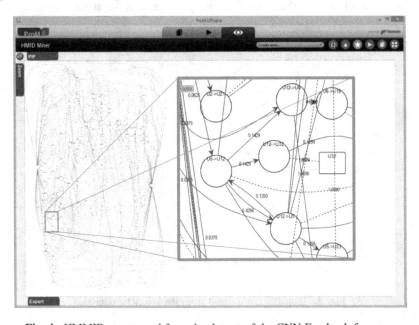

Fig. 4. HMMID constructed from the dataset of the CNN Facebook fan page.

The final Facebook dataset contains totally 80 posts, 22 actors, and 377 comments. The dataset generates 72 information diffusion states. For privacy reasons the name of actors were omitted and changed (e.g. Alfonso Gordillo to U12). The interaction, response, and probabilities of the information diffusion states are observable in Fig. 4. The communication pattern can be extracted from the flow of the information diffusion. The discovered HMMID reveals that the probability from 'U5 → U12' to 'U12 → U5' is high with a probability of 0.4286, and conversely the flow of information from 'U12 → U5' to 'U5 → U12' is lower with a probability of 0.1250. Therefore, is expected that, the flow of the information from 'U5 → U12' to 'U12 → U5' will be frequent.

Table 2. Observations in {*start*} state.

Rank	User	Probability %
1	U12	25.00
2	U5	23.75
3	U2	15.00
4	U21	11.25
5	U13	8.75
6	U8	6.25
7	U7	5.00
8	U19	3.75
9	U6	1.25

Table 2 shows the probabilities that different actors post a comment which starts the diffusion of the information inside the network. For this purpose, the information diffusion state {*start*} is analyzed, and then it is expected that U12 is the most important actor that helps to propagate initially the information on posts of CNN.

Figure 4 allows different analysis for information diffusion. Actors that have an influence to a specific actor can be analyzed. For example, examining actor U5 can be summarized the influence for U5 and observe the connections that exist with the actor, 'U21 → U5', 'U13 → U5', 'U12 → U5', 'U2 → U5', and so forth. In another analysis, the frequency of information diffusion states can be calculated. For example, analyzing the frequency of 'U21 → U13' can observe the influence from U21 to U13, to realize an analysis of the spread of information. Another analysis is to find the main paths that helps to propagate information between two given users, e.g. if the communication spread from U2 to U26 is needed to investigate, the main paths are <'U2 → U13', 'U13 → U26' > and < 'U2 → U12', 'U12 → U26'>.

6 Conclusions

The process discovery technique proposed in this research is used to better understand information diffusion in SNS and to represent the relationship among the users based on real-life SNS data. The proposed methodology, called HMMID, was developed by applying HMM technique to process mining. The HMMID miner was illustrate with a synthetic data and Facebook log data. The proposed methodology has an advantage of understanding the information flow more explicitly since the result diagram displays all possible sequences and paths of the information flow with the corresponding probabilities. Furthermore, the methodology is expected to be useful in various fields such as viral marketing and opinion mining because the diffusion process is analyzed with the probabilities that information spreads over possible information flow.

However, there are still a few drawbacks of this research. For example, although the results produce a proper information diffusion structure, it is hard to understand the relationship among users. Therefore more comprehensible social networking diagram can be developed to solve this problem. Also, the semantics of a specific information delivery cannot be obtained in this research because the keywords of comments are absent. These listed drawbacks will be discussed in further studies.

Acknowledgments. This work was supported by the National Research Foundation of Korea (NRF) grant funded by the Korea government (MSIP) (No. 2013R1A2A2A03014718).

References

1. Kwon, O., Wen, Y.: An empirical study of the factors affecting social network service use. Comput. Hum. Behav. **26**(2), 254–263 (2010)
2. Domingos, P.: Mining social network for viral marketing. IEEE Intell. Syst. **20**(1), 80–82 (2005)
3. Kim, H., Yoneki, E.: Influential neighbours selection for information diffusion in online social networks. In: 21st International Conference on Computer Communications and Networks, pp. 1–7. IEEE Press, Munich (2012)
4. Guille, A., Hacid, H., Favre, C., Zighed, D.: Information diffusion in online social networks: a survey. ACM SIGMOD Rec. **43**(4), 17–28 (2014)
5. Scott, J.: Social Network Analysis. SAGE, London (2012)
6. Duda, R.O., Hart, P.E., Stork, D.G.: Pattern Classification. Wiley, New York (2001)
7. van der Aalst, W.M.P.: Process Mining: Discovery Conformance and Enhancement of Business Processes. Springer, Heidelberg (2011)
8. Song, M., van der Aalst, W.M.P.: Towards comprehensive support for organizational mining. Decis. Support Syst. **46**(1), 300–317 (2008)
9. Kim, K., Jung, J.Y., Park, J.: Discovery of information diffusion process in social networks. IEICE Trans. Inf. Syst. **95**(5), 1539–1542 (2012)
10. Kim, K., Obregon, J., Jung, J.Y.: Analyzing information flow and context for facebook fan pages. IEICE Trans. Inf. Syst. **97**(4), 811–814 (2014)
11. Obregon, J., Jung, J.Y.: Discovering users-centric hierarchical process models in social networking services. In: Conference on Korean Institute of Industrial Engineering, pp. 480–489 (2014)
12. Bishop, C.M.: Pattern Recognition and Machine Learning. Springer, New York (2006)
13. Newman, M.: Networks: An Introduction. Oxford University Press, New York (2010)
14. Do, C.B., Batzoglou, S.: What is the expectation maximization algorithm? Nat. Biotechnol. **26**(8), 897–899 (2008)
15. Carrera, B., Jung, J.-Y.: Constructing probabilistic process models based on hidden markov models for resource allocation. In: Fournier, F., Mendling, J. (eds.) BPM 2014 Workshops. LNBIP, vol. 202, pp. 477–488. Springer, Heidelberg (2015)
16. Rabiner, L.R.: A tutorial on hidden markov models and selected applications in speech recognition. Proc. IEEE **77**, 257–286 (1989)

Process Visualization Techniques for Multi-perspective Process Comparisons

A. Pini[1]([⊠]), R. Brown[2], and M.T. Wynn[2]

[1] DensityDesign Research Lab, Politecnico di Milano, via Durando 10,
20158 Milan, Italy
azzurra.pini@polimi.it
[2] Business Process Management Discipline, Queensland University of Technology,
GPO Box 2434, Brisbane, QLD 4001, Australia
{r.brown,m.wynn}@qut.edu.au

Abstract. Organizations executing similar business processes need to understand the differences and similarities in activities performed across work environments. Presently, research interest is directed towards the potential of visualization for the display of process models, to support users in their analysis tasks. Although recent literature in process mining and comparison provide several methods and algorithms to perform process and log comparison, few contributions explore novel visualization techniques. This paper analyzes process comparison from a design perspective, providing some practical visualization techniques as analysis solutions. In order to support the needs of business analysts the design of the visual comparison has been tackled via three different points of view: the general model, the superimposed model and the side-by-side comparison. A case study is presented showing a preliminary evaluation of the application of process mining and visualization techniques to patient treatment across two Australian hospitals.

Keywords: Comparative visualization · Process mining · Business process management

1 Introduction

Recently, the necessity of managing and analyzing a large number of processes together with their growing complexity has brought an increasing interest towards methods and technologies to support the representation and comparison of process models. The comparison activity might need to focus, for example, on the discrepancies between the real behavior as captured by event logs and the reference process model, the analysis of process variants to understand the differences, and even cross-organizational process comparisons to describe the peculiar characteristics of each system and to identify the best practices for process improvement. Process mining [2] a research domain formed by combining data mining and process analysis techniques, has developed techniques to

J. Bae et al. (Eds.): AP-BPM 2015, LNBIP 219, pp. 183–197, 2015.
DOI: 10.1007/978-3-319-19509-4_14

analyze processes by making use of event logs. Nevertheless, within the reference literature related to process mining and business process management, the visualization dimension of comparison is still in an exploratory stage and there is a demand to elaborate effective solutions to facilitate this activity for both process analysts and stakeholders.

At the same time, we note the availability of a broad and deep corpus of research in the information visualization field, containing techniques generally not applied to business process data, resulting in a lack of specific contributions exploring the aspect of visualization for process comparison [4,24]. Research has shown the superior utility of visual representations as compared to table data [24] and we argue that the intended audience for this research, business analysts, cognate about business systems from a control flow perspective, so processes should be represented as a graph of temporally ordered activities shown to match their internal model of the business [28]. In addition, there are a number of perspectives to these processes; control, resource and data [3,12], that need to be understood clearly by the analyst to improve the process aligned with the model.

One of the ways to better understand how to improve business processes of an organization is to compare the behavior and performance of processes within the organization against others who are carrying out similar kinds of operations. Process variants represent alternative ways of performing business activities to accomplish a goal. It is important to understand the reasons for these variations as well as the effects of such variations on process performance in order to make process improvement recommendations. Regrettably, this potential is not fully realized yet, as the majority of existing process mining techniques analyze a single log at a time and this step then needs to be repeated for all the process variants of interest [10,14,16]. As a result, the comparison between the behavioural and performance aspects of different process variants is carried out by manually (and potentially subjectively) interpreting the results.

This concept of process benchmarking or learning from the results of other similar processes in businesses is a well-accepted notion in business process management [7,20], which will be applied in this paper. We motivate our multi-perspective approach in this paper by noting that since particular analysis tasks are aligned with these perspectives [1,12], any visualization approach, sensitive to these requirements [27], should be able to visualize all these perspectives effectively. Thus, this paper proposes new comparative process visualization techniques which combine approaches from process management and information visualization fields to communicate the similarities and differences between the behavior and performance of business processes.

In Sect. 2, an analysis of related work is presented. In Sect. 3 we present a series of techniques designed for comparative process visualization to assess performance and behavior differences among various cohorts. The new techniques engage the comparative perspective through three different views: *general model*, *superimposed model* and *side-by-side comparison* with the ultimate goal of extracting indications for process improvement. Section 5 continues with a description of the preliminary evaluation we have performed, including example visualizations created using real hospital process data and the feedback from presentation to hospital management stakeholders. Section 6 concludes the paper.

2   Related Work

Our exploration stems from two streams of BPM research, process visualization and process comparison. At the same time our research refers to more general concepts belonging mainly to information visualization, in order to find possible intersections and useful techniques applicable to multi-perspective process comparison.

As more and more organizations rely heavily on IT systems to support their business operations, a vast amount of detailed records of business operations (i.e., which activities are carried out by whom at what time for which customer and at which cost) becomes available for analysis. Sophisticated process mining techniques can be applied to this data in order to reveal the real behaviour and performance of these operations [2]. While visualization techniques have been widely recognized as crucial for supporting decision making and analysis tasks as well as the emergence of behavioral patterns [13,24,29,31], within BPM we register just a few relevant contributions with an interest in visualization aspects, especially regarding personalized views [26], process change [15,32] and dynamic visualization [19].

A number of papers recently explored aspects related to process comparison in different ways. Kleiner [14] analyzed the technique of delta analysis for comparing the actual process represented by a process model with some reference process considered as a prescriptive process model. Delta analysis provides a basis for process comparison by generating a similarity measure between the reference and the discovered process models by using an estimate of the equivalence of event logs. The analysis though is performed only from the data perspective and does not focus on the implementation of a graphical model to show the control flow perspective. The time dimension also emerges from process visualization literature as particularly significant for process data [4,19]. Although processes are intrinsically characterized by the time dimension, process modeling has rarely visualized it. Currently, the only time structure that is represented in process graphs is the ordering of activities as a workflow sequence, without any indication of duration of activities or waiting time between them.

A number of contributions concerning the relationship between several process variants with a reference or general model have emerged in the literature. Küster [17] focuses on the consolidation of process models though the automatic detection and resolution of differences between process versions. Li, Reichert and Wombacher [20] concentrate their analysis on the minimization of the derived reference model from a set of process variants. Process similarity has also been studied by Dijkman [10] mostly in terms of metrics and search algorithms for business process model repositories, focusing on model structure and behavior similarity metrics. The technique can be usefully applied to the computation of a difference map, which together with a side-by-side arrangement, represents the main approach to process comparison. While the second one mainly relies on the user's visual memory to operate the comparison by pulling the models alongside, the difference map consists in computing a merged model summarizing the differences and similarities of the compared processes. A few contributions consider though the two different approaches together [5, 15].

A number of papers also tackle the aspect of comparison of process variants with graphical representations, using mostly the color variable to represent the differences across both activities and links [9, 15, 16, 21]. Most of the visualization approaches perform the comparison only on two process models, using color-coding to present the difference analysis as a comparison to a reference model, referring to differences as "deletion","addition" and "changes". As a consequence they always use one of the two processes as a reference to operate the comparison. Focusing on instance traffic, Kriglstein et al. [16] explore a number of visualization techniques to compare process models. A difference analysis is performed between two process models and the visualization specifies the discrepancies on activities and edges through a color-coding approach. A more appropriate approach [15] has been adopted by the same authors for the visualization of changes in business processes to highlight the intermediate steps that lead to an updated process. Andrews [5] instead presents a semantic graph visualizer to calculate and visualize the similarity of graph components. The approach applies a difference map visualization by associating a color to each graph, merging the two hues in a gradient for common nodes in a difference map. A different color-coding has been applied by Buijs [7] for a dual comparison visualization of process models and their executable logs. The alignment matrix visualization proved to be too complex and difficult for participants to interpret.

We also explored contributions outside BPM and information systems disciplines, such as graph drawing [6] and information visualization [13, 29]. The field of uncertainty visualization has also investigated the representation of similarity measures [8, 22]. The literature review has highlighted the lack of significant disciplinary connections between the fields of BPM and process mining and the information visualization disciplines, suggesting a need for a design approach to guide the development of novel visualization techniques to support process comparison activities.

3   Comparative Visualization Design Approach

Process mining is a well-established research discipline that exploits event data using a combination of process analysis and data mining techniques [2]. Using process mining techniques, one can automatically discover a process model (and related resource usage and performance metrics) from an event log [2]. However, in order to carry out a comparative analysis of processes, process mining techniques are first applied to a single log (optionally with a single process model) and this step is then repeated for all processes of interest. As a result, the comparison between behavioural and performance aspects of different processes is then carried out by manually interpreting the results. As most existing process mining techniques do not cater for comparisons in an automated and straight-forward manner, there are also challenges in making use of existing visualizations from process mining frameworks and tools such as ProM[1] or Disco[2]. Hence, there is a real need for novel comparative visualizations that can highlight key differences in terms of process behaviour and performance.

4   Data Requirements

In this paper, we address the key requirements in process analysis to be able to visualize the differences in terms of process behaviour and performance of two or more processes, while making use of different process-related information including process models and historical records of process executions. We identify the three main inputs to the visualizations: logs, process models and visualization configurations.

The main input for the proposed visualization solutions is one or more event logs. The event log(s) are used to extract data regarding process behaviour and performance. In particular, the information regarding the set of completed activities, the frequency of those activity executions and the min/max/avg duration of those activities will be used as objective measures for the visualizations. An event log could be as minimal as having only one transition type (i.e., "complete" events). With richer logs such as those with start and complete timestamps, additional customer information or employee data, it is possible to have a more accurate picture regarding wait times, bottlenecks and resource utilizations.

Furthermore, our proposed visualization solutions heavily rely on the existence of one or more process models to map performance differences upon or to compare and contrast different ways of executing processes. The process models are used to visualize the order in which activities are being carried out. It is, of course, possible to use the input event log(s) to discover these process models using existing process discovery techniques [2]. In theory, any process modelling language (e.g., BPMN, Petri Nets, EPC or Fuzzy model) can be supported.

The final input is the desired visualization configurations which enable the selection of data streams (logs) and related process models and mapping of data to generate relevant visualizations. Thus, this overview of techniques should be seen within the context of a complete interactive system for manipulation of process mining data for comparison purposes, providing the ability to obtain an overview, drill down and compare models as required [29]. In Sect. 5 we show an example where a visualization is configured and viewed for hospital data case studies. We now proceed to describe in detail the design of these visualizations from basic principles.

4.1   Visualization Techniques

The requirements analyzed in the previous section have motivated some design examples to tackle representational issues in process comparison. Design

---

[1] Process Mining Tools - http://www.promtools.org/.
[2] Disco - http://www.fluxicon.com/disco/.

solutions were developed for cohort comparison in general, in one single organization or across multiple organizations. The comparison has been tackled from different perspectives in order to capture the different aspects of variability in the processes. In order to bridge some of the gaps identified in the literature we directed design efforts to the different perspectives of process mining, in particular the time, performance and resource perspectives. Concerning the comparative perspective we consider the possibility to comparing more than two models. Although the comparison of multiple models has already been explored in [7, 25, 30], none of the analyzed contributions examines the design of an actual difference model that considers the characteristics of all compared models.

The proposed visualization techniques have been conceived to allow the exploration both globally as an overview and individually on the single processes, supporting the user moving across different abstraction levels [24]. All three views have been designed aiming at comparing processes both at the model level and event log data, in order to include information regarding the performance, time and resource perspectives. Each view is complementary to the others, focusing on different process mining perspectives and users' points of view, in order to highlight varied aspects of comparison. The proposed examples visualize the comparison across the process models for three cohorts, which are identified by three different color hues (red, blue and green).

General Model. The aim of the general model view is to observe the differences between cohorts with a focus on the differences in the performance and resource perspectives (Fig. 1). The starting point of the visualization is one process model which represents the general process model for the different cohorts.

In order to illustrate our approach, we consider the three main attributes that represent the basic components associated to activity execution, that is *activity name, median duration* and *frequency*, in addition to the number of resources (see Table 1). The values for median duration and frequency for each cohort are normalized on each activity proportionally for each cohort, to obtain performance related data. Next, for each activity, resources are aggregated per organizational level across the different cohorts, in order to display the ratio between performance (frequency/duration) and number and type of resources involved in each activity (see activity A in Fig. 1). Resources have been classified into three organizational levels for explanatory reasons, following a typical hierarchy of managerial, professional and technical staff. The examples thus indicate the different resources levels performing the particular task, shown by circles with differing color fills (refer to the left part of Fig. 2).

The example visualization in Fig. 1 applies the stacked bar pattern (described below) for highlighting the differences in performance of the cohorts in each activity. For each activity a stacked colored bar is partitioned according to the different execution time of each cohort. Color transparency is used to map activity frequency, assigning a higher alpha value to a lower frequency.

Different visual patterns, by way of glyphs (see Fig. 2), have been explored for the representation of performance variations across different cohorts at the

Table 1. Sample data attributes used in the visualizations

Activity	Performance (P)		Resources			Similarity
	median duration	frequency	level 1	level 2	level 3	
Injury	5.52'	112	0	2	1	0.5
...	...	...	...	...	...	...

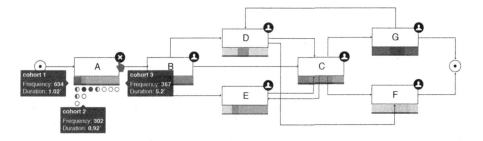

Fig. 1. General model example, with merged model and log data annotations.

activity level. In each case, the different blocks of color represents a different cohort performing the activities. The stacked bar (Fig. 2a), applied also in the example in Fig. 1, constitutes an immediate way to map the differences across activities directly to the model, obtaining both an analytic and global view. By implementing multiple color dimensions, other information such as the absolute frequency of each activity can be mapped within the stacked bar, allowing for comparison across other processes. In order to maintain readability, color transparency has been rendered through a range of four different non-continuous levels.

A similar alternative for the representation of this data type is a space filling visualization of hierarchies, such as a treemap representation (Fig. 2b). Keeping the hue variable associated to cohort categorical values and transparency to map frequency data, the performance/temporal value is represented on the space (area), providing more uniformity in case of a high variability in values. A different solution applies overlapping circle sections for each cohort (Fig. 2c), by mapping the frequency to the radius and the median duration on the arc section subtended angle. This solution have been designed to stress the difference between cohorts and to represent the time dimension as a percentage of the maximum completion time. An overlapping principle has also been explored through triangle shapes (Fig. 2d) associated to each cohort. This allows a mapping of the performance values, i.e. frequency and median duration, to height and base width respectively. This type of pattern might be more appropriate for models that are not particularly complex, when the design goal is to perform a comparison at the activity level than the control-flow one. At the same time it might reveal some issues in readability in case performance data is too similar, causing the superimposed triangles to overlap. For particularly complex models,

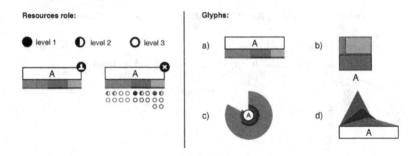

Fig. 2. Visualization of the resource perspective and associated glyphs for the general model shown in Fig. 1

a more suitable solution is to concentrate on the control-flow perspective and eliminate all possible sources of visual occlusion, thus delegating the comparative perspective to single activities with interaction elements that can be activated and deactivated whenever necessary, as displayed in Fig. 1. We are currently working on an online survey with the goal of deeply assessing the strengths and weaknesses both of the three views and of the different visual patterns.

Superimposed Model. The superimposed model view is devoted to the comparison of different cohorts following the perspective of one process model, that we identify here with the first cohort (C1). The main aspect for consideration is the correspondence of activities in the model, visualized through the alignment and superposition of an activity element as in [11].

The main aspects considered in the different cohorts are the process flow (i.e. activity ordering) and the similarity of activities. The similarity level of activities can be based on different values depending on the aspect to be observed, varying from unidimensional factors such as execution time and frequency to metrics modelling the general performance. The example presented in Table 1 and Fig. 3 considers similarity in terms of the ratio of cohorts performance values, between the frequency and the median duration of each activity in C1 compared to the average value of the same ratio for C2 and C3. The resulting values are grouped

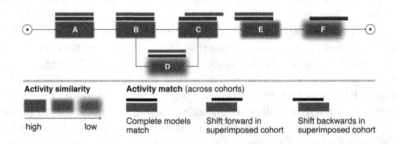

Fig. 3. Visualization of superimposed cohorts: C2 and C3 - over C1

by level of similarity in three partitions: high, medium and low. The similarity scores of activities in C1, with respect to C2 and C3, is mapped by applying three levels of blur as in [8], according to the partitions, where the highest level of blur corresponds to the lowest similarity level for the activity across the cohorts. The superimposition of the models is based on the match of activity position within the process flow across the different cohorts. The presence of each activity is checked in the three models, as well as its direct predecessors and successors, to verify if the same activity is executed in different parts of the process, thus establishing the presence of a shift in the ordering of activity execution, forward or backwards. The matching activities are mapped as a stacked rectangles on the top of the reference process model (C1). The rectangle is then slightly shifted towards the left when the same activity is founded in the model but in a different position, earlier in the flow, and towards the right when the same activity occurs at some point later in the flow.

Side-by-side Model. This type of comparison technique aims at exploring, more deeply, the time perspective of the processes at a broader level, by integrating the information on the waiting time between an activity and its successor: a very common event that causes the delay of completion times for the whole process.

The three models are analyzed separately, focusing specifically on the ordering of activities. The proposed diagram (Fig. 4) exploits the process model logical flow to describe temporal dependencies between activities through predecessor and successor nodes of a directed graph [23]. In order to capture the variability across the models we applied a visualization approach that highlights just the matching flows that correspond to the comparison scenario, leaving the irrelevant branches in the background [8]. This approach requires a further analysis of log data. Besides the main properties used for the general and superimposed

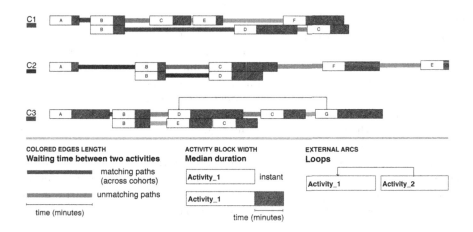

Fig. 4. Side by side model comparison

model, the information related to the waiting time is extracted and stored in a separate source/target table, identifying the couples of consecutive activities. The waiting time between each couple of activities is represented by the length of the arcs, while activity duration is displayed by extending the activity box with a grey texture. This visualization method is also consistent with a configurable process model approach [18]. This type of comparison might present some issues in case of particularly complex models. Especially if the models present a large variability in the waiting time between activities, further calculations are required in terms of data normalization, in order to maintain the readability of the diagram.

5   Evaluation

The evaluation approach adopted for the proposed visualization framework is three-fold. Firstly, we made use of event logs and discovered process models from two hospitals (H1 and H2)[3] and developed a set of visualizations by hand. This serves as a preliminary evaluation and feasibility analysis of the proposed design principles. Secondly, we showed the resulting visualizations to the stakeholders in order to (1) gauge the understandability and usefulness of proposed visualizations and (2) to solicit further user requirements.

Finally, we are in the process of developing a set of software plug-ins for the process mining framework, ProM, based on their input and are also preparing an anonymous online survey in order to obtain the opinions of BPM practitioners and academics from around the world. In this paper, we present the evaluation outcomes from the first two steps: visualizations created using real datasets and stakeholder feedback about the visualizations. Please note that due to the lack of resource information in the datasets, the visualizations do not include the resource perspective.

Hospital One. One of the comparative analysis questions from stakeholders at Hospital One (H1) is "Are there any differences in terms of process behaviour and durations for patients who present at ED at different times of the day?" In order to answer this question, patients are put into four cohorts depending on their arrival times at ED (i.e., midnight - 6am, 6am - 12noon, 12noon - 6pm and 6pm - midnight). A process model, together with dominant paths, is discovered from the event log containing data for all four cohorts. The names of the activities, their frequencies and median execution times of activities are calculated for each cohort.

Figure 5 depicts the resulting visualization. From this figure, it is easy to see the performance comparison across two dimensions (frequency and duration) for four different cohorts. As the number of cases for each cohort varies across the different time periods (i.e. 147, 244, 320, 173 min), the relative frequencies

---

[3] These event logs represent patients presenting for treatments at Emergency Departments (ED) of two QLD hospitals.

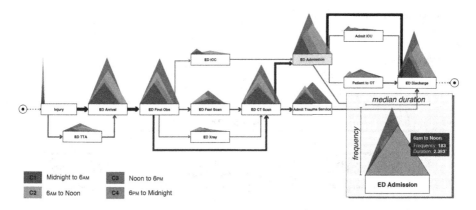

Fig. 5. Visualizing the behaviour and performance differences between four patient cohorts in H1. The ED Admission activity is blown up on the bottom right.

are used in the visualization. The visualization made use of a number of metric classes: the absolute frequency for activities (the height of the triangles), the absolute frequency for paths (the strength of the edges and activity darkness), and the median duration from one activity to another (the width of the triangles).

One example of a pattern being easily seen is the difference in the ED Admission activity for the 6am to Noon cohort, compared to the others, shown by the wide triangle indicating a large difference in duration compared to other cohorts (see highlighted box bottom right in Fig. 5). As this was the first visualization created with the real data sets, further refinements to the original design were necessary. For instance, we needed to adjust the dimensions of the visualization elements in order to accommodate very high/low frequencies. We also realized that it might be necessary to set the maximum limit with respect to the number of cohorts being compared. This visualization was presented to stakeholders (including doctors from the emergency department at H1, as well as healthcare researchers from different QLD Hospitals) as a part of three presentations to demonstrate preliminary results from the process mining analysis being conducted at H1. These stakeholders found the visualization to be intuitive and they were very receptive to being presented with visual comparisons of the four cohorts across the two performance dimensions.

Hospital Two. Another comparative analysis question from stakeholders at Hospital Two (H2) is "What are the differences in terms of process behaviour and durations for patients who are discharged from ED within four hours of arrival and those who stayed longer than four hours?" In order to answer this question, the dataset is split into two cohorts, those who stayed in ED for less than or equal to four hours and those who stayed for more than four hours. All three types of visualizations were created using the data from H2. Process models were created for both cohorts as well. For this evaluation we concentrate

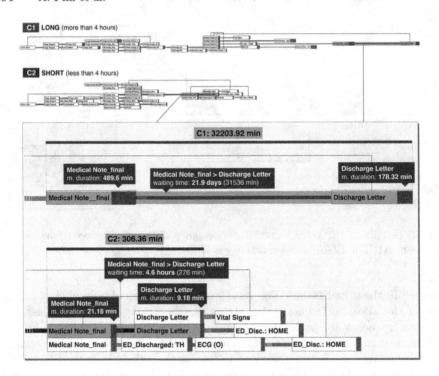

Fig. 6. Side-by-side comparison of two patient cohorts in H2, with a blown up selected example at the bottom.

on the superimposition and side by side visuals as the performance general model visualization is similar to the example for H1.

Figure 6 depicts the visualization that reflects the side-by-side comparison of patients in the two cohorts. Here, the emphasis is on the time perspective whereby cases in Cohort One (C1) have throughput time of up to 4 hours and cases in Cohort Two (C2) has throughput time of over 4 hours. The design also allows the comparison of dissimilar models by the selection of two similar segments of the H1 and H2 models for comparison. As seen in the example the portion of process between Medical Note_final and Discharge Letter is significantly longer in C1, due to the waiting time as well as the median duration of both activities involved.

Figure 7 depicts the superimposition of the process model for C1 onto the model for C2 with emphasis on whether the activities are being shifted forward or backward in relation to a model. The example shows that the activity related to ECG (ordered) is executed later in the model in C2 with respect to C1, while Medical Note final has the same position in both cohorts but with a lover level of similarity, as displayed by the blur.

These two visualizations were shown to the head of the emergency department from H2. This doctor found all three visualizations to be useful for different

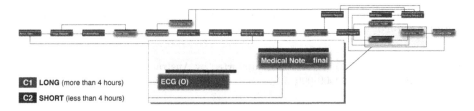

Fig. 7. Superimposed model of two patient cohorts in H2, with a blown up selected example at the bottom.

purposes. He noted that the performance models (e.g., Fig. 5) provide salient patterns that pop-out easily. Figure 6, showing time-based visuals using alignment analysis, was seen as useful as it highlighted the differences in time easily, seeing activities related to particular antecedents. Figure 7, which highlights the differences between the process behaviour of the two cohorts was found to be sub-optimal for this dataset due to a high degree of similarities found across the two cohorts; thus minimal blurring. However, he recognized the potential use of this type of visualization in comparing different departments or different hospitals with a high level of variation in process behaviour.

Findings from these preliminary evaluations also highlight the need for an integrated system starting at a high level, filtering and drilling down to activity comparisons, with interactions assisting with insight in real time. We are currently working on a software plug-in to support these visualizations with interaction and filtering capabilities.

6   Conclusion

In this paper we have presented research on a collection of multi-perspective visualization techniques for process comparisons. These designs emerge from a need to better communicate process comparisons within the process mining domain. Our research has highlighted the lack of design approaches for comparative process visualizations, and the scarcity of efforts in visual patterns innovation for the representation of processes. In particular, we developed a design approach to tackle representational issues within process comparison activities and a series of display techniques for comparing multiple cohorts across four perspectives, namely: control-flow, time, performance and resources. The evaluation phase drew attention towards the positive response of stakeholders in respect to experimentation on visual patterns in process representation, as well as in the availability of different views to address different process perspectives. As a general objective, we intend to continue to broaden the research in process visualization and search for improvements in the visual patterns and interaction modes for process mining analysis activities. For future work, we plan to work on the implementation of the proposed visual solution within a dynamic environment, such as ProM. We also aim to expand the evaluation of the visualizations with a systematic survey to assess the effectiveness of the different representations.

Acknowledgement. This research is supported by Australian Centre for Health Services Innovation (#SG00009-000450).

References

1. van der Aalst, W.M.P., Netjes, M., Reijers, H.A.: Supporting the full BPM life-cycle using process mining and intelligent redesign. In: Siau, K. (ed.) Contemporary Issues in Database Design and Information Systems Development, ch. 4, IGI Global, Hershey, USA, pp. 100–132 (2007)
2. van der Aalst, W.: Process Mining: Discovery, Conformance and Enhancement of Business Processes. Springer, Heidelberg (2011)
3. van der Aalst, W., ter Hofstede, A., Kiepuszewski, B., Barros, A.: Workflow patterns. Distrib. Parallel Databases **14**(1), 5–51 (2003)
4. Aigner, W., Miksch, S., Muller, W., Schumann, H., Tominski, C.: Visualizing time-oriented data: a systematic view. Comput. Graph. **31**(3), 401–409 (2007)
5. Andrews, K., Wohlfahrt, M., Wurzinger, G.: Visual graph comparison. In: 13th International Conference Information Visualisation, pp. 62–67 (2009)
6. Archambault, D.: Structural differences between two graphs through hierarchies. In: Proceedings of Graphics Interface 2009 (2009)
7. Buijs, J.C.A.M., Reijers, H.A.: Comparing business process variants using models and event logs. In: Bider, I., Gaaloul, K., Krogstie, J., Nurcan, S., Proper, H.A., Schmidt, R., Soffer, P. (eds.) BPMDS 2014 and EMMSAD 2014. LNBIP, vol. 175, pp. 154–168. Springer, Heidelberg (2014)
8. Collins, C., Carpendale, S., Penn, G.: Visualization of uncertainty in lattices to support decision-making. In: Proceedings of the 9th Joint Eurographics/IEEE VGTC Conference on Visualization, pp. 51–58 (2007)
9. Cordes, C., Vogelgesang, T., Appelrath, H.-J.: A generic approach for calculating and visualizing differences between process models in multidimensional process mining. In: Fournier, F., Mendling, J. (eds.) BPM 2014 Workshops. LNBIP, vol. 202, pp. 383–394. Springer, Heidelberg (2015)
10. Dijkman, R., Dumas, M., van Dongen, B., Käärik, R., Mendling, J.: Similarity of business process models: metrics and evaluation. Inf. Syst. **36**(2), 498–516 (2011)
11. Gleicher, M., Albers, D., Walker, R., Jusufi, I., Hansen, C.D., Roberts, J.C.: Visual comparison for information visualization. Inf. Vis. **10**(4), 289–309 (2011)
12. Jablonski, S., Goetz, M.: Perspective oriented business process visualization. In: ter Hofstede, A.H.M., Benatallah, B., Paik, H.-Y. (eds.) BPM Workshops 2007. LNCS, vol. 4928, pp. 144–155. Springer, Heidelberg (2008)
13. Keim, D.: Information visualization and visual data mining. IEEE Trans. Vis. Comput. Graphics **8**(1), 1–8 (2002)
14. Kleiner, N.: Delta analysis with workflow logs: aligning business process prescriptions and their reality. Requirements Eng. **10**(3), 212–222 (2005)
15. Kriglstein, S., Rinderle-Ma, S.: Change visualization in business processes - requirements analysis. In: VISIGRAPP/IVAPP. Rome, Italy (2012)
16. Kriglstein, S., Wallner, G., Rinderle-Ma, S.: A visualization approach for difference analysis of process models and instance traffic. In: Daniel, F., Wang, J., Weber, B. (eds.) BPM 2013. LNCS, vol. 8094, pp. 219–226. Springer, Heidelberg (2013)

17. Küster, J.M., Gerth, C., Förster, A., Engels, G.: Detecting and resolving process model differences in the absence of a change log. In: Dumas, M., Reichert, M., Shan, M.-C. (eds.) BPM 2008. LNCS, vol. 5240, pp. 244–260. Springer, Heidelberg (2008)

18. La Rosa, M., Dumas, M., Ter Hofstede, A.H., Mendling, J.: Configurable multi-perspective business process models. Inf. Syst. **36**(2), 313–340 (2011)

19. de Leoni, M., van der Aalst, W.M.P., ter Hofstede, A.H.M.: Visual support for work assignment in process-aware information systems. In: Dumas, M., Reichert, M., Shan, M.-C. (eds.) BPM 2008. LNCS, vol. 5240, pp. 67–83. Springer, Heidelberg (2008)

20. Li, C., Reichert, M., Wombacher, A.: Discovering process reference models from process variants using clustering techniques. Technical report TR-CTIT-08-30, Enschede, March 2008

21. Lin, Y., Gray, J., Jouault, F.: Dsmdiff: a differentiation tool for domain-specific models. Eur. J. Inf. Syst. **16**(4), 349–361 (2007)

22. MacEachren, A.M., Robinson, A., Hopper, S., Gardner, S., Murray, R., Gahegan, M., Hetzler, E.: Visualizing geospatial information uncertainty: what we know and what we need to know. Cartography geographic Inf. Sci. **32**(3), 139–160 (2005)

23. Mendling, J., Simon, C.: Business process design by view integration. In: Eder, J., Dustdar, S. (eds.) BPM Workshops 2006. LNCS, vol. 4103, pp. 55–64. Springer, Heidelberg (2006)

24. Moody, D.L.: The "physics" of notations: toward a scientific basis for constructing visual notations in software engineering. IEEE Trans. Software Eng. **35**(6), 756–779 (2009)

25. Partington, A., Wynn, M., Suriadi, S., Ouyang, C., Karnon, J.: Process mining for clinical processes: a comparative analysis of four australian hospitals. ACM Trans. Manage. Inf. Syst. **9**(4), 1–18 (2013)

26. Reichert, M., Kolb, J., Bobrik, R., Bauer, T.: Enabling personalized visualization of large business processes through parameterizable views. In: SAC 2012, EE 2012, pp. 1653–1660 (2012)

27. Sedlmair, M., Meyer, M., Munzner, T.: Design study methodology: reflections from the trenches and the stacks. IEEE Trans. Vis. Comput. Graphics **18**(12), 2431–2440 (2012)

28. Shaft, T.M., Vessey, I.: The role of cognitive fit in the relationship between software comprehension and modification. MIS Q. **30**(1), 29–55 (2006)

29. Shneiderman, B.: The eyes have it: a task by data type taxonomy for information visualizations. In: IEEE Symposium on Visual Languages, pp. 336–343 (1996)

30. Suriadi, S., Mans, R.S., Wynn, M.T., Partington, A., Karnon, J.: Measuring patient flow variations: a cross-organisational process mining approach. In: Ouyang, C., Jung, J.-Y. (eds.) AP-BPM 2014. LNBIP, vol. 181, pp. 43–58. Springer, Heidelberg (2014)

31. Tominski, C., Forsell, C., Johansson, J.: Interaction support for visual comparison inspired by natural behavior. IEEE Trans. Vis. Comput. Graphics **18**(12), 2719–2728 (2012)

32. Weber, B., Reichert, M., Rinderle-Ma, S.: Change patterns and change support features: enhancing flexibility in process-aware information systems. Data Knowl. Eng. **66**(3), 438–466 (2008)

Author Index

Printed in the United States
By Bookmasters